Remediation of Uranium Mill Tailings

This book covers issues pertaining to uranium tailings with special reference to consolidation of radioactivity including systematic ecological strategy for consolidation of radionuclides in uranium tailings. It discusses sustainable consolidation of radioactivity and checks the migration of unextracted uranium from tailing piles to plants and atmosphere supported by a case study from a uranium mine. It provides simple ecological solutions for the remediation of radioactivity in mill tailings.

Features:

- Provides insight into the application of applied ecology for bioremediation of radioactive wastes.
- Discusses species selection criteria for tailings radioactivity consolidation.
- Explains safe treatment of the tailings of radioactive ore processing plants.
- Illustrates the role of ethnobotany in the selection of the most appropriate species to effectively use in bioremediation.
- Focuses on experimental outcomes.

This book is aimed at researchers and professionals in mining engineering, applied geology, nuclear tailings and environmental protection.

Remediation of Uranium Mill Tailings

Prafulla Soni

CRC Press
Taylor & Francis Group
Boca Raton London New York

CRC Press is an imprint of the
Taylor & Francis Group, an **informa** business

Designed cover image: Prafulla Soni

First edition published 2024
by CRC Press
6000 Broken Sound Parkway NW, Suite 300, Boca Raton, FL 33487–2742

and by CRC Press
4 Park Square, Milton Park, Abingdon, Oxon, OX14 4RN

CRC Press is an imprint of Taylor & Francis Group, LLC

© 2024 Prafulla Soni

ISBN: 9781032352794 (hbk)
ISBN: 9781032352800 (pbk)
ISBN: 9781003326151 (ebk)

DOI: 10.1201/9781003326151

Typeset in Times
by Apex CoVantage, LLC

Contents

About the Author

Prafulla Soni is a forest ecologist with more than 50 years of dedicated research experience in the field of restoration ecology biodiversity and environmental conservation of degraded landscapes. She has trained in watershed management and environmental impact assessment in forests. Through the years, she has been involved in the planning, execution and coordination of research and consultancy projects on ecology and environment with particular reference to ecological restoration of derelict landscapes.

More than 20 students have successfully completed research under her supervision on various aspects of forest ecology, including restoration technologies developed for opencast mined areas based on the ecosystem approach. These technologies incorporate the development of a stable, self-sustaining system using ecologically and socioeconomically viable plant species that assist in ecological maintenance of essential processes, restoration of biodiversity and sustainable utilization of species and ecosystems. The technology has been transferred to different mining companies, according to their site-specific requirements, through consultancy and capacity-building initiatives for senior executives and government officials.

To share the field and research experiences, she was invited by the US Department of the Interior to represent her country in international symposia on degraded lands and has visited France, the Netherlands, the US and Australasian countries as a representative of the Forest Research Institute/Government of India. Prafulla has 140 research papers, four authored/edited books and various reports to her credit and is a life member of the Society of Indian Foresters, and the Indian Association of Soil and Water Conservationists. She is founder member, Fellow and president of the Society for Applied Forest Ecological Research (SAFER) and editor-in-chief of e-Journal *Applied Forest Ecology* (eJAFE).

Preface

The movement and subsequent off-site dispersal of radionuclides are essential factors for risk assessments and determining remedial actions at hazardous, radioactive waste sites. One potential transport path from the soil column is through the uptake by vegetation and transport to surface soil. The most significant radionuclides that may appear in the environmental matrices are U (nat) ^{226}Ra and polonium in view of the waste fixation technology.

mines in India are operated by the Uranium Corporation of India (UCIL), which is owned by the Atomic Energy Department of Government of India, and are located in the tribal belt of Jharkhand. Generation and subsequent disposal of tailings in the tailing ponds have exerted a significant impact by fragmenting the natural forest landscape. These tailings, although very low in radioactivity, need to be restore dinan ecologically safe manner so that the plant species used to restore the tailings may both consolidate the radioactivity and pose no health hazard to people living in the area and their live stock.

Although the recorded radioactivity level in these tailings was very low, to avoid any long-term impact on the atmosphere and biosphere, a case study sponsored by the Board of Research in Nuclear Studies (BRNS) was undertaken to discover the appropriate plant species of forestry origin for phytostabilizing the tailings and consolidating radioactivity below the soil surface.

A rapid ecological appraisal of the uranium mines and tailing pondsat Jaduguda was taken up. A floristic survey and species richness study in the adjoining natural forests and tailing ponds were carried out.

Most plants take up their nutrients and radionuclides mainly via the roots from the soil. Revegetation is essentially required to consolidate the radioactivity in the tailings so that it does not emanate into the atmosphere or enter the food chain. Uptake and distribution of radionuclides in soil capping, plant roots and shoots are in tailings>soil capping on tailings>roots of selected plant species>shoots of all the selected species.

To avoid translocation of radionuclides through the food chain, plant species with low root penetration, with low crown cover, of no use by the local population, that can grow well on soil-capped tailings and that persist indefinitely with little or no maintenance are recommended.

The book has been designed in a way that gives insight into uranium mining globally, the issues and challenges of mining and milling operations. The radionuclide uptake and dissemination in the soil and plants and their impact on human body are discussed. The book would prove useful for students of environmental management and activists practicing restoration.

Acknowledgments

The author is thankful to the Board of Research in Nuclear Sciences (BRNS), Government of India for providing financial support to initiate remediation of uranium tailings. Research and scientific support of Dr. Lal Singh, senior scientist CSIR-NEERI, deserves special mention, without whom this project, particularly the case study, would not have been possible. Dr. V.D. Puranic, Dr. (Mrs.) Susan Eappan, and Dr. Venu Babu from Bhabha Atomic Research Center (BARC), Mumbai, deserve special thanks for their support.

1 Introduction

As long as uranium ore is undisturbed deep underground, not much radioactivity in the form of radon diffuses to the surface. But when uranium ore is brought to the surface, radon is released into the atmosphere where it can be inhaled. Uranium extraction requires the need of an oxidant for effective leaching, particularly in the acid leach process. The addition of chemicals may enhance the pollution potential of the processed waste. Due attention is required to minimize the risk due to the presence of chemical additives in the waste. Manganese and iron have important bearing in the uranium industry. Manganese is added in the process, whereas iron is abundant in the ore body itself. These two elements are interrelated in their mode of bioavailability. Both elements are soluble under a low pH environment and consequently are mobilized. In a high pH environment, they are mostly insoluble and mobility is inhibited to a great extent. Iron is precipitated even at a lower pH. Rankama and Sahama (1950a) have extensively studied the geochemistry of these two elements. The hazard associated with excessive intake of manganese (manganese poisoning) has been reviewed by Venkatraman (1981).

Uranium tailings are generated as solid and liquid wastes in uranium mining/milling operations. Since most of the uranium deposits in the world have low grades, millions of tonnes of such wastes are produced annually. Often, the uranium tailings are locally disposed of, using sites with suitable conditions to construct tailings basins. The main concern during the operation of a disposal site is the presence of radium in the liquid phase that overflows from dewatered tailings. This radium is precipitated by adding a proper chemical reagent. The potential environmental hazards of uranium tailings arise when the disposal site is abandoned after the decommissioning of the uranium mill. Huge amounts of solid waste as small particles of depleted ore remain in place. Topsoils are usually for soil stabilization and for controlling radon emissions. Hence, water infiltration through uranium tailings presents a potential hazard to underground aquifers. Water plays a dual role by triggering a sequence of reactions and by carrying contaminants away from the waste sites. The situation is more serious in a country like Canada, where the precipitation rate is higher than the evaporation rate and the water table is about one meter beneath the ground surface. For permanent disposal of uranium tailings, either the hazardous constituents of these wastes have to be isolated from the percolating water or the subsoils must have acceptable sealing and sorptive characteristics. The first approach is affected by blending proper solidifying reagents with the tailings to totally block them from the leaching effects of aggressive pore water. Solid waste treatment may prove costly in additive usage and the blending operation. The leaching of toxic constituents, however, becomes effective as long as the tailings are in a loose form. This is because of the acid-generating properties of the pyrite, a substantial constituent (upto 8% wt) of the tailings. Two types of contaminants tend to leave the solid and migrate to the

DOI: 10.1201/9781003326151-1

pore water. These are the heavy metals and the radioactive elements, both of which are hazardous for their toxicity and persistence.

Natural uranium (U) occurs as a mixture of three radioactive isotopes: 238U, 235U and 234U. Only 235U is fissionable, and it makes up about 0.7% of natural U, while 238U is overwhelmingly the most abundant at greater than 99% of the total mass of U. Prior to the 1940s, U was predominantly used as a coloring agent, and U-bearing ores were mined mainly for their radium (Ra) and/or vanadium (V) content; the bulk of the U was discarded with the tailings (Finch et al., 1972). Once nuclear fission was discovered, the economic importance of U increased greatly. The mining and milling of U-bearing ores is the first step in the nuclear fuel cycle, and the contact of residual waste with natural water is a potential source of contamination of U and associated elements to the environment.

Accelerating demands of energy have necessitated at least nuclear energy enhancement to the tune of 25% for electricity generation during the next three decades as part of a clean and reliable low-carbon mix. Achieving this means that nuclear generation must triple globally.

1.1 DISTRIBUTION OF URANIUM MINES ACROSS THE GLOBE AND THEIR GRADING

In the last five to six decades, uranium has become one of the most important minerals for energy generation. Most of it is used to generate electricity, but a small amount is used for medical and other purposes.

Globally, uranium occurs naturally in the outermost layer of the earth with an average concentration of 2.8 parts per million, although it occurs almost everywhere and in all rocks and soils. According to the World Nuclear Association (2021), the total identified uranium resource available globally so far is 5.5 million metric tonnes, while it is presumed that an additional 10.5 million metric tonnes still remain unexplored. Further, nearly 4.5 billion tonnes of uranium is available in seawater, but at present it is not economically feasible to extract it. It is estimated that 5.5 million tonnes of uranium exists in ore reserves that are economically viable for extraction, while 35 million tonnes are classed as mineral resources with logical prospects for final economic extraction depending on technological advances for recovery. However, unlike other metals, concentration of uranium is very low. Most uranium mines across the globe usually produce very low-grade ores containing 0.1% to 0.3% U_3O_8. As high as120 ppm is found in phosphate rocks of Florida and southeastern Idaho. The neighboring areas can be used as commercial sources of uranium (Clegg and Foley, 1958; NCRP, 1975). The high uranium content of phosphate rocks is reflected in corresponding elevated levels in phosphate fertilizers.

Recoverable resources that are reasonably assured, as well as inferred resources, have been listed by OECD-NEA and IAEA (2020). The total recoverable identified resources to $260/kg U is8.070 million tonnes U. Commercial-grade uranium can be produced through the reduction of uranium halides with alkali or alkaline earth metals. Uranium metal can also be prepared through electrolysis of KU_5orUF_4, dissolved in molten

TABLE 1.1

Uranium Production from Mines during the Last Decade (in Tonnes)

S.No	Country	2011	2012	2013	2014	2015	2016	2017	2018	2019	2020
1.	Kazakhstan	19,451	21,317	22,451	23,127	23,607	24,689	23,321	21,705	22,808	19,477
2.	Australia	5983	6991	6350	5001	5654	6315	5882	6517	6613	6203
3.	Namibia	3258	4495	4323	3255	2993	3654	4224	5525	5476	5413
4.	Canada	9145	8999	9331	9134	13,325	14,039	13,116	7001	6938	3885
5.	Uzbekistan (est.)	2500	2400	2400	2400	2385	3325	3400	3450	3500	3500
6.	Niger	4351	4667	4518	4057	4116	3479	3449	2911	2983	2991
7.	Russia	2993	2872	3135	2990	3055	3004	2917	2904	2911	2846
8.	China (est.)	885	1500	1500	1500	1616	1616	1692	1885	1885	1885
9.	Ukraine	890	960	922	926	1200	808	707	790	800	744
10.	India (est.)	400	385	385	385	385	385	421	423	308	400
11.	South Africa (est.)	582	465	531	573	393	490	308	346	346	250
12.	Iran (est.)	0	0	0	0	38	0	40	71	71	71
13.	Pakistan (est.)	45	45	45	45	45	45	45	45	45	45
14.	Brazil	265	326	192	55	40	44	0	0	0	15
15.	USA	1537	1596	1792	1919	1256	1125	940	582	58	6
16.	Czech Republic	229	228	215	193	155	138	0	0	0	0
17.	Romania	77	90	77	77	77	50	0	0	0	0
18.	France	6	3	5	3	2	0	0	0	0	0
19.	Germany	51	50	27	33	0	0	0	0	0	0
20.	Malawi	846	1101	1132	369	0	0	0	0	0	0
21.	Total world	53,493	58,493	59,331	56,041	60,304	63,207	60,514	54,154	54,742	47,731
22.	tonnes U$_3$O$_8$	63,082	68,974	69,966	66,087	71,113	74,357	71,361	63,861	64,554	56,287
23.	% of world demand	87%	94%	91%	85%	98%	96%	93%	80%	81%	74%

Source: Data based on the World Nuclear Association (Anonymous, 2016–2022). Updated in September 2021.

calcium chloride ($CaCl_2$) and sodium chloride (NaCl) solution. Very pure uranium is produced through the thermal decomposition of uranium halides on a hot filament.

There have been experiments to extract uranium from seawater, but the yield has been low due to the carbonate present in the water.

About 96% of the global uranium reserves are found in ten countries: Australia, Canada, Kazakhstan, South Africa, Brazil, Namibia, Uzbekistan, the United States, Niger and Russia (Anon, 2022). According to the World Nuclear Association (Anonymous, 2016–2022), during 2020 Kazakhstan produced the largest share of uranium from mines (41% of world supply), followed by Australia (13%) and Canada (8%). Table 1.1 shows a detailed account of the distribution and production of uranium. Kazakhstan has about 12% of the world's reserves, or about 651 kilo tonnes (1.4×10^9 lb) and Canada has 485 kilotonnes ($1,100 \times 10^6$ lb) of uranium, representing about 9% (Wikipedia, 2022).

The top ten largest producing uranium mines produce 51% of total uranium (Table 1.2). Cigar Lake Canada is the highest producer with 3885 tonnes of uranium. Australia has large reserves of uranium with significant deposits in northern Australia.

Mining companies usually consider concentrations greater than 0.075% (750 ppm) as ore, or rock economical to mine at current uranium market prices, and grading is done accordingly (World Nuclear Association, 2008).

TABLE 1.2
The Largest-Producing Uranium Mines in 2020

Mine	Country	Main owner	Type	Production (tonnes U)	% of world
Cigar Lake	Canada	Cameco/Orano	underground	3885	8
Husab	Namibia	Swakop Uranium (CGN)	open pit	3302	7
Olympic Dam	Australia	BHP Billiton	by-product/ underground	3062	6
Inkai, sites 1–3	Kazakhstan	Kazatomprom/ Cameco	ISL	2693	6
Karatau (Budenovskoye 2)	Kazakhstan	Uranium One/ Kazatomprom	ISL	2460	5
Rössing	Namibia	Rio Tinto	open pit	2111	4
SOMAIR	Niger	Orano	open pit	1879	4
Four Mile	Australia	Quasar	ISL	1806	4
South Inkai (Block 4)	Kazakhstan	Uranium One/ Kazatomprom	ISL	1509	3
Kharasan 1	Kazakhstan	Kazatomprom/ Uranium One	ISL	1455	3
Top 10 total		**24,162**	**51%**		

1.2 ISSUES RELATED TO URANIUM MINING AND FUTURE PROSPECTS

The mining of uranium has been controversial mainly because of its associated impacts not only on the environment but also on human health. The conventional way of exploiting the rock ores—surface mining, heap leaching or underground mining—aim formaximum production to fulfill the escalating demand, which has led to major environmental, economic and social risks. According to Srivastava et al. (2020), technological advancement in mining practices could significantly reduce some impacts compared to the early strategic period, and the inevitable water-intensive extraction process is a serious concern, as is the handling of mine tailings, which contain thorium, radium, uranium and polonium and emit radon.

Though the major portion of uranium present in the ore is extracted, a fraction remains unextracted and is finally discharged with tailings. The unextracted fraction may be either soluble hexavalent complexes of uranium or the insoluble tetravalent form. After the ore is processed, the tailings that remain are released in tailing ponds specially designed for this purpose. These tailings still have some radioactivity, which although they are very low in radioactivity, they need to be restored in an ecologically safe manner to avoid long-term adverse effect on the environment. The radioactivity in tailings must be consolidated so that it does not emanate into the atmosphere or enter the food chain. Radiological exposure in the tailing ponds is usually negligible as the waste had very low level of activity concentration.

While energy generation in power plants by using fossil fuels results in severe land degradation, nuclear energy has a very low physical footprint and a negligible amount of destruction of forests and helps to preserve biodiversity (World Nuclear Association, 2021). Although at a uranium mine ordinary operating procedures normally ensure that there is no significant water or air pollution, typical environmental problems arising from mill tailings include radon emanation, windblown dust dispersal, and the leaching of contaminants, including radionuclides, heavy metals and arsenic, into surface and ground waters. Radon (Rn) emissions are due to exhalation from the waste materials, and the Rn can reach the ambient atmosphere when free circulation of air in the material and its cover is possible. Emissions to water bodies occur when infiltration of precipitation is unhindered, bottom liners are absent and no collection of drainage waters is installed. The leaching of contaminants is usually exacerbated by acid formation from pyrite oxidation under conditions of varying degrees of saturation with water. Additional effects from acid rain have also been observed. In many instances, contaminants other than radionuclides may be the real problem. Regardless of how uranium is removed from rock, the extraction process creates radioactive wastes. If not managed properly, mining waste and mill tailings can lead to severe environmental contamination.

Uranium eventually decays to radium, and radium decays to release a radioactive gas called radon. Open pit uranium milling and *in situ* mining sites do not pose a significant radon risk to the public or to miners, but it may be harmful when the radon disperses into the atmosphere. In the past, the waste rock produced by underground and open pit mining was piled up outside the mine. This practice has caused problems. Wind can blow radioactive dust from the wastes into populated areas, and the wastes can contaminate the surface water used for drinking or irrigation. Some sites also have considerable groundwater contamination.

Underground mines can present a radiation hazard to miners. Without proper air ventilation, radon can collect in the mineshafts, where it is inhaled by miners. The operators of uranium mines must take special precautions to protect miners, such as pumping radon gas out of the mine and replacing it with fresh air. To protect the public near uranium mines, vented radon gas must not exceed certain limits.

Previously, waste rock and mill tailings were used in some Western mining areas as building materials for homes, schools, roads and other construction. Structures built with waste rock and mill tailings had radon and radiation hazards to anyone spending time in them. People traveling on roads made with waste rock were in danger of breathing radioactive dust. In response to these issues, the Uranium Mill Tailings Radiation Control Act (UMTRCA, 1978) stopped the use of mill tailings in building and construction projects.

The movement and subsequent off-site dispersal of radionuclides are essential factors for risk assessments and determining remedial actions at hazardous, radioactive waste sites. One potential transport path from the soil column is through the uptake by vegetation and transport to surface soil. The behavior and mobility of radionuclides in soil is a major consideration for plant uptake and is influenced by many variables (Romney et al., 1974 and Poston et al., 1995a). The uptake characteristics of radionuclides have been observed to vary depending on soil properties such as texture, organic matter content, bacterial action, pH, redox potential and physiochemical speciation (Poston et al., 1995b; White and Dunaway, 1975; Romney et al., 1974; Han and Lee, 1997; Salbu and Oughton, 1995). Naturally occurring uranium consists of three isotopes with mass numbers 238, 235 and 234. Uranium-238 constitutes 99.28% of natural uranium, which is in equilibrium with ^{234}U (0.0058%). Uranium-235, the progenitor of the actinium series, is present in the amount 0.71% (Eisenbud, 1987). In the uranium ore, which is not subjected to leaching, radionuclides are expected to be in secular equilibrium. For 95% average recovery of uranium in the processing plants, about 14% of the total radioactivity contained in the ore remains in the uranium concentrate.

With regard to environmental security, long-lived radionuclides of ^{238}U decay series are the prime cause of concern viz. ^{230}Th, ^{226}Ra, ^{210}Pb and ^{210}Po. They have half-lives of 80,000 years, 1622 years, 22 years and 138 days respectively. Owing to the low abundance of ^{235}U in the ore (0.7%), its daughter products are of less significance for environmental protection.

The two most significant radionuclides that may appear in the environmental matrices are U (nat) and ^{226}Ra in view of the waste fixation technology. The former is chemically more toxic and accumulate sin the kidney, whereas the latter is a bone seeker with behavior similar to calcium (a known essential element for bone development). Consequently, from the point of view of abundance, established chemical/radiological toxicity and long half-lives, these two elements are also the most significant radionuclides in the uranium processing industry.

During the extraction of uranium ore and its processing, some releases of uranium occur. The releases of radium and other decay products of uranium are normally more important than the uranium in tailing ponds at the mines and ore processing centers (Anon, 2012). The nature of wastes in the form of uranium mill tailings is unique, primarily due to their significant radioactivity. Although milling removes about 90% of the contained uranium, the decay products are retained in the tailings, comprising some 85% of the original radioactivity. The principal radionuclides of concern are thorium-230 and radium-226, with half-lives of about 75,000 and 1600 years, respectively. Each element has different chemical properties, thus providing a complex challenge in managing the radioactive tailings in perpetuity. The decay product of radium-226 is radon-222, a noble gas with highly radioactive decay products (Mudd, 2000).

According to Waggitt (1994), uranium ore often contains higher concentrations of heavy metals and trace elements, giving rise to the potential for chemical toxicity from surface water or groundwater contamination. Gupta and Sarangi (2005) have emphasized that in Indian uranium mines, worldwide technological progress in the field of ore processing has been kept in mind and resulted in absorbing expertise and adapting cutting-edge technology through radical innovations. A great deal of effort has already been made to implement precipitation of uranium peroxide ($UO_4.2H_2O$) using hydrogen peroxide in place of magnesium di uranate. This will prevent co-precipitation of other metals, ensure higher purity in product and control many environment-related problems.

Rehabilitated mill tailing deposits may be physically and chemically secure, but without proper attention given to radionuclide behavior, excessive gamma radiation and radon emanation may give rise to higher radiological exposure risks. The abandonment of uranium mines initially developed in the 1950s and 1960s are examples of major environmental insult (Bastias, 1987; Ryan, 1987). Another abandoned uranium mine, that at Rum Jungle in the Northern Territory, was responsible for the costliest rehabilitation program ever undertaken on an Australian mine (approximately A\$20 m). Uranium and copper ore concentrates were produced between 1954 and 1971 from five open cuts in close proximity. After mine closure, no rehabilitation was required or undertaken (Ryan, 1987). The mining and processing operations resulted in the pollution of the nearby East Fitness River with acid mine drainage resulting from oxidation of pyrite in the mine waste and overburden and causing elevated levels of aluminum iron, copper, zinc, cobalt and manganese.

1.3 MILLING AND PROCESSING OF URANIUM ORE

At present, about 46% of uranium comes from conventional mines (open pit and underground), about 50% from *in situ* leach, and 4% is recovered as a by-product from other mineral extraction. Mining methods keep on changing. Underground mines produced 55% of world production until 1990, but by 1999 this reduced radically to 33%, but again by 2000 the new Canadian mines increased it. Solid wastes at uranium mill tailing (UMT) sites can include both standard tailings (i.e., leached ore rock residues) and solids generated on site by waste treatment processes. The latter can include sludge or "mud" from neutralization of acidic mine/mill effluents, containing Fe and a range of coprecipitated constituents, or barium sulfate precipitates that selectively remove Ra (Campbell et al., 2015). Underground mines produced

TABLE 1.3

Production of Uranium by Different Extraction Methods

Method	tonnes U	%
In situ leach (ISL)	26,402	55
Underground & open pit (except Olympic Dam)	18,017	38
By-product	3312	7

55% of world production until1990, but by 1999 this reduced radically to 33% but again by 2000 the new Canadian mines increased it. Uranium production by *in situ* leaching process accounted for 55% of global production while underground and open pit mines accounted for only 38% and production as by-product was only 7%. *In situ* leach (ISL, also called *in situ* recovery, ISR) mining has been steadily increasing its share of the total, mainly due to Kazakhstan, and in 2020 accounted for over half of production (Table 1.3).

The 95th United States Congress enacted the Uranium Mill Tailings Radiation Control Act (UMTRCA, 1978) and 2017 as Public Law, codified at 42 USC §7901 *et seq.* Congress found that uranium mill tailings

> may pose a potential and significant radiation health hazard to the public, and that every reasonable effort should be made to provide for stabilization, disposal, and control in a safe, and environmentally sound manner of such tailings in order to prevent or minimize other environmental hazards from such tailings.

Congress authorized the secretary of energy to designate inactive uranium processing sites for remedial action by the US Department of Energy (DOE). Congress also directed the US Environmental Protection Agency (EPA) to set the standards to be followed by the DOE for this process of stabilization, disposal and control.

In mining, raw uranium ore is brought to the surface and crushed into a fine sand. Uranium tailings are therefore a waste by-product (tailings) of uranium mining. The valuable uranium-bearing minerals are then mechanically removed, and the remaining radioactive sand, called "uranium tailings", is stored in huge impoundments.

These uranium tailings are left on the surface where, if not properly attended to, drying occurs. This dry radioactive sand can be carried vast distances with wind, entering the food we band water bodies. Although the majority of tailing mass will be in the form of inert rock, similar to the raw ore (before the extraction of the uranium), during the process of mining followed by the generation of tailings, the mass is not only physically altered by the process of grinding, mixing with large amounts of water and being exposed to atmospheric oxygen, but also may be substantially altered in chemical behavior.

Uranium tailings contain over a dozen radioactive nuclides that are the cause of exposure and risk. The most important of the two radionuclides that are potentially hazardous are thorium-230, radium-226, radon-222 (radon gas) and the daughter isotopes of radon decay, including polonium-210. Radionuclides contained in uranium tailings emit 20 to 100 times as much gamma radiation as natural background levels on deposit surfaces. Gamma radiation levels decrease rapidly with distance from the pile.

The radium-226 in tailings continuously decays to the radioactive gasradon-222, the decay products of which can cause lung cancer. Some of this radon escapes from the interior of the pile. Radon releases are a major hazard that continues long after uranium mines stop extraction and processing. In the early 1990s, concerns were raised about the stability and environmental performance of tailing dams and impoundments. To mitigate these concerns, such embankments often rely on a certain amount of controlled seepage to enhance stability, which in turn affects environmental performance. Ritcey (1989) has speculated that the need for sound impoundments in the uranium industry "probably" accounts for much of the recent attention paid to impoundment design in other types of facilities. Perhaps triggered by the initial attention to uranium impoundments, the increasing concern for environmental performance has led to better engineering design of tailing dams in other mining industry sectors, for both stability and environmental performance. For instance, experience gained with leach pad liners is being transferred to linings for tailing ponds, and the use of synthetic lining materials is growing (although use of liners is still far from being the industry norm). Since radon spreads rapidly with wind, people staying in the vicinity regularly receive small radiation doses, which is a cause of concern.

Another problem with tailings is that after rainfall, gullies may form leading to seepage and infiltration of water. Seepage poses risk of contamination to ground and surface water. Residents in the vicinity may also be affected by radium-226 and other hazardous substances like arsenic in their drinking water supplies and in fish from the area. The seepage problem is very important with acidic tailings, as the radionuclides involved are more mobile under acidic conditions. In tailings containing pyrite, acidic conditions automatically develop due to the inherent production of sulfuric acid, which increases migration of contaminants to the environment. Plants and burrowing animals can penetrate into the tailing deposit, leading to dispersal of tailing material and thus enhancing the radon emanation and making the deposit more susceptible.

1.4 OVERBURDEN AND TAILINGS MANAGEMENT AND THEIR RESTORATION

Every mining activity leaves behind impacts on the different components of the environment—be it soil, water, air or society. Uranium mining is no different but rather has more critical impacts in the inherent radioactivity and therefore contamination with radioactive dust, radon gas, water-borne toxins and increased levels of background radiation. These radioactive wastes generated during mining in the form of either mining waste or mill tailings lead to higher environmental contamination. Most uranium mines in the world usually produce low-grade ores containing 0.1% to 0.3% U_3O_8 (Table 1.4). Mines in India have still lower grades than these. Therefore, the uranium industry generates large quantities of waste. Almost the entire ore mined comes out as waste after recovery of the small amounts of uranium that it contained, contaminating large areas with radioactive waste.

TABLE 1.4

Grades of Uranium According to Concentration in the Ore

Source	Concentration
Very high-grade ore—20% U	200,000 ppm U
High-grade ore—2% U	20,000 ppm U
Low-grade ore—0.1% U	1,000 ppm U
Very low-grade ore—0.01% U	100 ppm U
Granite	4–5 ppm U
Sedimentary rock	2 ppm U
Earth's continental crust (av)	2.8 ppm U
Seawater	0.003 ppm U

In open pit mining, overburden (nonmineralized soil/rock that covers an ore body) and waste rock (poorly mineralized or very low-grade soil and rock that are within the ore body or surrounding it) are first removed from the entire area of the final planned pit, often in very large quantities. The mine shape is then formed by a series of benches and terraces arranged in a deepening spiral or in levels with interconnecting ramps. The pit is then deepened in a sequential manner using benches that also serve as haulage roads for the removal of ore from the pit. Rehabilitation at the end of the mining process is usually very expensive, particularly if it involves backfilling, as the wastes are often dumped at some distance to avoid obstructing the removal of economic ore. In open cast mining, on the other hand, extraction proceeds laterally rather than vertically, and this method is therefore more suitable for shallow and laterally extensive deposits. A "strip" of overburden is first removed to expose the underlying economic ore deposit, where after the deposit is then worked out. Rehabilitation normally takes place as part of the mining process and also involves backfilling, revegetation and maintenance in a sustainable manner.

During the early phases of uranium mining and milling, due importance was not given for safe disposal of the waste in some countries. This resulted in indiscriminate discharge of uranium mill tailings into large water bodies. Solid wastes were also used as landfills and building materials. Subsequently, the radioactivity level in the recipient waters increased to alarming levels. River basins were contaminated with sediment rich in radioactivity. This activity finally could reach humans through different biosphere pathways. These problems have since been recognized, and in recent years great emphasis has been laid on the method of treatment of uranium mill wastes. Safe confinement on a long-term basis and controlled releases into the environment at acceptable levels were ensured. At present, strict regulatory control is being practiced for safe management of tailings and associated discharges (Markose, 1990).

Countries like Australia and Canada have stringent regulations according to which they are usually required to follow numerous environmental laws. However, many countries don't have these regulations. This often leads to enormous problems due to both the toxicity of uranium and the mining methods used. Indian uranium deposits are of medium size, and the country has a modest uranium resource. Only a small part of the land mass of the total of 3.28 million sq km of the Indian subcontinent is assumed to be geologically favorable for hosting uranium deposits. This results in the generation of a large volume of solid waste and effluent. With greater public awareness of health hazards and stringent environmental guidelines, the management of these tailings (solid and liquid waste) has become a crucial part of the emergent uranium mining sector. As more and more low-grade ore is required to be processed in the future to recover uranium, the technological breakthrough through unconventional ways of extraction should gradually be put into practice. The possibility of adopting low-cost, large-volume uranium ore processing through heap leaching in low-grade ore of Singhbhum and Tummalapalle may be looked into at the appropriate time. A cluster of small tonnage deposits also need to be considered for commercial exploitation, though production of uranium on a small scale does not in any way reduce the inherent problems of uranium mining and processing.

The uranium tailings management system needs to be strengthened and to demonstrate best practices, which warrants wide-ranging research and development. Newer technologies to bring in sound design features of the embankment system and impermeability of the pond surface need to be developed. Restoration and/or appropriate land use management techniques are needed to ensure that the health and safety of the affected population are not compromised.

Globally there is need for judicious management of uranium mining and milling waste. Assessments sponsored by IAEA of the radiological conditions at mines and mills in several countries have shown that uranium production can be carried out in a safe manner without detriment to the environment. A list of IAEA publications in the field of environmental restoration is given in Table 1.5.

The IAEA report (1992) discusses the current practices used in the design, siting, construction and operation of impoundment facilities for uranium mill tailings.

Since most of the uranium deposits in the world have low grades, millions of tonnes of such wastes are produced annually. Although nearly all of the uranium is extracted from the mined material, the tailings are not less hazardous (a large amount of the contaminants of the total radioactivity and all the associated chemical contaminants are still present), and therefore dumping the tailings back to the excavation site from where the ore has been taken from does not result in an acceptable solution for disposal. The chances of migration of these hazardous substances into the environment persist. In most cases, uranium mill tailings are disposed in some form or another to limit release of contaminants into the environment.

Uranium tailings are generated as solid and liquid wastes in uranium mining/milling operations. Tailings dams are often not of stable construction. In most cases, they were made from sedimentation of the coarse fraction of the tailings sludge. Some are built on geologic faults, therefore, they are subject to the risk of failure due to strong rain or snow storms. The main concern during the operation of a disposal site is the presence of radium in the liquid phase that overflows from dewatered tailings. This radium is precipitated by adding a proper chemical reagent. The potential environmental hazards of uranium tailings arise when the disposal site is abandoned after the decommissioning of the uranium mill. Huge amounts of solid

TABLE 1.5

List of IAEA Publications in the Field of Environmental Restoration

Safety Series (SS)	
1. Safe management of wastes from the mining and milling of uranium and thorium ores: Code of practice and guide to the code	SS No. 85 (1987)
Technical Reports Series (TRS)	
1. Cleanup of Large Areas Contaminated as a Result of a Nuclear Accident	TRS No. 300 (1989)
2. The Environmental Behaviour of Radium	TRS No. 310 (1990, 2 vols.)
3. Planning for Cleanup of Large Areas Contaminated as a Result of a Nuclear Accident	TRS No. 327 (1991)
4. Disposal of Wastes From the Cleanup of Large Areas Contaminated as a Result of a Nuclear Accident	TRS No. 330 (1992)
5. Measurement and Calculation of Radon Releases from Uranium Mill Tailings	TRS No. 333 (1992)
6. Current Practices for the Management and Confinement of Uranium Mill Tailings (Supersedes TRS No. 209)	TRS No. 335 (1992)
7. Decommissioning of Facilities for Mining and Milling of Radioactive Ores and Closeout of Residues	TRS No. 362 (1994)
Technical Documents (TECDOC)	
1. Environmental Migration of Radium and Other Contaminants Present in Liquid and Solid Wastes from the Mining and Milling of Uranium	IAEA-TECDOC 370 (1986)
2. In-Situ Leaching of Uranium: Technical, Environmental and Economic Aspects	IAEA-TECDOC 492 (1989)
3. Uranium In-Situ Leaching	IAEA-TECDOC 720 (1993)
4. Planning and Management of Uranium Mine and Mill Closures	IAEA-TECDOC-824 (1995)
5. Planning for Environmental Restoration of Radioactively Contaminated Sites in Central and Eastern Europe	IAEA-TECDOC 865 (1996, 3 Vols.)
6. Closeout of Uranium Mines and Mills: A Review of Current Practices	IAEA-TECDOC-939 (1997)
Proceedings Series	
1. Management of Wastes from Uranium Mining and Milling. IAEA-OECD/NEA Symposium, Albuquerque, USA, 1982	Proceedings Series

waste as small particles of depleted ore remain in place. Topsoils are usually for soil stabilization and for controlling radon emissions. Hence, water infiltration through uranium tailings presents a potential hazard to underground aquifers. Water plays a dual role by triggering a sequence of reactions and by carrying contaminants away from the waste sites. The situation is more serious in a country like Canada, where the precipitation rate is higher than the evaporation rate and the water table is about one meter beneath the ground surface. For permanent disposal of uranium tailings, either the hazardous constituents of these wastes have to be isolated from the percolating water, or the subsoils must have acceptable sealing and sorptive characteristics. The first approach is affected by blending proper solidifying reagents with the tailings to totally block them from the leaching effects of aggressive pore water. Solid waste treatment may prove costly in additive usage and the blending operation. The leaching of toxic constituents, however, becomes effective as long as the tailings are in a loose form. This is because of the acid-generating properties of the pyrite, a substantial constituent (up to 8% wt) of the tailings. Two types of contaminants tend to leave the solid and migrate to the pore water. These are the heavy metals and the radioactive elements, both of which are hazardous for their toxicity and persistence.

The situation is similar for deposit of tailings in former open pit mines. Here also, immediate contact to groundwater exists, or seepage presents risks of contamination of groundwater. Only in the case of the presence of proven impermeable geologic or human-made layers can the contamination risk to groundwater be prevented.

In France and Canada, the concept of dumping the tailings in former open pits in ground water has been pursued or proposed at several sites in recent years. In this case, a highly permeable layer is installed around the tailings, to allow free groundwater circulation around the tailings. Since the permeability of the tailings themselves is lower, it is anticipated (by the proponents) that nearly no exchange of contaminants between tailings and groundwater takes place. A similar method is being tested in Canada for the disposal of uranium mill tailings in lakes.

Conventional mining of uranium is similar to mining other metalliferous ores, and hence there are substantial volumes of barren rock and overburden waste. These are placed near the pit and either used in restoration or shaped and revegetated. Mining methods, tailings and run-off management and land rehabilitation are subject to government regulation and inspection in different countries. For instance, in Australia the Code of Practice and Safety Guide: *Radiation Protection and Radioactive Waste Management in Mining and Mineral Processing*, was published in 2005.Once mining is over, tailings are covered permanently with enough clay and soil to reduce both gamma radiation levels and radon emanation rates to levels near those naturally occurring in the region and enough rock to resist erosion. A vegetation cover is then established. Since environmental restoration is a multidisciplinary activity, an integrated approach to optimize resources has to be developed. Also, it should be noted that radioactive contamination is part of the overall environmental contamination in the region (IAEA, 1997). Wastes generated during mining and milling processes need to be restored in an ecologically safe manner so that the plant species used to restore the tailings do not pose any health hazards to people living in the area or to their livestock. Appropriate screening of the plant species used to take up any amount of radionuclides is required so that the plants are not used by people or cattle. This will check the movement of radionuclides in the food chain. Further, this would help the mining sector and the locals, as well as the site ecology. It is therefore stressed that sustainable development of the mined areas with an ecosystem approach is essential, instead of simple revegetation or bioreclamation. Eco rejuvenation of these areas based on sound ecological principles is important (Wali, 1987, Soni et al., 1994).

India has almost very meager resources of the uranium. The grade of the uranium ore dug in these uranium mines is remarkably low. The average grade is about 0.06%. The smelter in Jaduguda can only process 1000 tonnes of ore per day[2]. And the amount of uranium is only 600 kg per day, namely 200 tonnes per year. In the meantime, mill tailings after uranium is removed reaches 300,000 tonnes per year. They have been thrown into tailing ponds by pipeline in liquid form. Besides, the mine tailings also occur, and the quantity reaches 10 times of the mill tailings. They are left in surroundings without any countermeasures. The main component of natural uranium is U-238, and its half-life is 4.5 billion years, which is almost equal to the age of the earth. U-238 itself is a radioactive nuclide, and the generated nuclides are also radioactive and repeat to decay one after another. It finally becomes lead (Pb-206) after changing 14 kinds of radioactive nuclides. Those radionuclides are called "daughters" or "progenies". As they decay, some daughter nuclides such as radium and radon pose a unique danger.

Uranium was reported to be the most frequent radionuclide contaminant in groundwater and surface soils in more than 50% of DOE facilities involved in nuclear operations (Riley et al., 1992). Generally, the tailings are covered with a relatively impermeable layer (hydraulic conductivity of 10^{-9} to 10^{-8} m s^{-1}) of waste rocks, overburden, clay and soil cover. In many countries, the total thickness of the cover material is of the order of 1 to 2 meters. This may, however, be site specific but should aim at reducing the gamma radiation to 0.2 Gy.h^{-1} above background and radon emanation rate to 0.74 Bqm^{-2} s^{-1}(IAEA, 1987) or as approved by the regulatory body.

1.5 RADIOACTIVITY CONSOLIDATION IN MINE WASTE

Uranium is regarded as the heaviest trace element found in nature, and it occurs (Kirk-Othmer, 1969) in numerous minerals. Generally, the tailings are covered with a relatively impermeable layer (hydraulic conductivity of 10^{-9} to 10^{-8} m s^{-1}) of waste rocks, overburden, clay and soil cover. In many countries the total thickness of the cover material is of the order of 1 to 2 meters. This may, however, be site specific.

While geomechanical aspects, such as the stability of pile slopes, dikes and retaining dams, are standard engineering problems, for which in most countries provisions are made in the relevant building or mining regulations, environmental and radiological impacts have often been neglected. It should be mentioned, however, that mill tailings as such can pose serious engineering challenges, owing to the geomechanical and physicochemical characteristics of the sediments. Environmental problems from mill tailings are radon emanation, windblown dust dispersal, and the leaching of contaminants, including radionuclides, heavy metals and arsenic, into surface and groundwaters. Radon (Rn) emissions are due to exhalation from the waste materials, and the Rn can reach the ambient atmosphere when free circulation of air in the material and its cover is possible. Emissions to water bodies may also occur if infiltration of precipitation is unhindered and bottom liners are absent with no system for collection of drainage waters is installed (IAEA, 2004a).

A variety of technical measures can be employed to prevent or reduce the extent of these processes. Capping can be used to control radon emanation, moisture infiltration and chemical reactions that may promote leaching. The physical and chemical properties of the tailings can be improved *in situ* or by reprocessing to enhance long-term stability. Containment structures can be improved to meet the minimum factor of safety. However, any of these engineering solutions have a finite life span, and hence ecological solutions may provide sustainable solutions.

Plants can be used to consolidate radioactivity in tailings from the soil, either as an economic resource or to consolidate contamination. In the UK, trials have demonstrated that some plants from the cress genera have the ability to extract zinc, cadmium and nickel from the soils. When these plants are harvested and incinerated, their ash can contain up to 20% metal oxides that can be removed by conventional methods. This mechanism is so effective that, once reclaimed, the land should be suitable for agriculture. In California, the same approach is being tested, aiming at recovering nickel and cobalt from the soil with the aid of the wild mustard plant. Scientists expect the ash of the plants, sown on nickel-bearing soils, to contain between 15% and 20% nickel, which they will then recover using standard leaching procedures.

In the early 1970s, the first major research program into developing the requirements and methods to establish vegetation on sulphide tailings was started. This research proved the necessity of including legumes in tailings revegetation programs. The superiority of bird's-foot trefoil (*Lotus corniculatus* L.) as the legume for this purpose was established. Saskatchewan is the largest producer of uranium in Canada, with an annual production of 7722 tonnes per year. One of the largest uranium mills in the world is located at Key Lake, Saskatchewan.

A special study on vegetative covers was recently carried out as part of the Uranium Mill Tailings Remedial Action Project (UMTRAP) in the United States (US Department of Energy, 1988). This study investigated the use of vegetation to stabilize uranium tailings and specifically includes the use of vegetation to intercept infiltration. The principal finding of the study is that properly developed plant communities on complex soil covers can be effective in stabilizing covers and controlling infiltration on top slopes of waste piles. The study showed that the appropriate vegetative cover will adapt to climatic change, will repair itself after severe disturbances such as fires and drought and will persist indefinitely with little or no maintenance. The plants were found to protect top slopes against sheet wash erosion; however, resistance to gully erosion depends more on the overall pile configuration than on the vegetation and soil.

Certain physical, chemical and vegetative stabilization methods have been evaluated for purpose of mine waste reclamation by the US Bureau of Mines (Dean et al., 1986). The techniques for managing uranium tailings are essentially the same as those for managing tailings from the mining of any other mineral. The fact that uranium tailings are moderately radioactive does not alter their tailings management requirements. Occasionally, because of fine sandy texture, dried tailings have been used for house construction or for landfills. High levels of gamma radiation and radon were found in such houses built on or from such material. The US EPA has estimated a higher lung cancer risk of residents of such houses.

2 Ecological Strategies for Consolidation

Ecological strategy for consolidating radio active waste is basically managing and treating all solid and liquid waste in a manner such that no living or non-living component of the environment is affected by any nuclear traces. Ecologically compatible vegetation or plants that show negligible uptake of radionuclides and trace elements are best suited for consolidating radioactivity for short periods of time over the tailings due to the probability of radionuclides and trace metals from reaching the food chain. Due to their shallow root penetration and survival under harsh conditions, they are attractive candidates for plantation over the remediated and soil-covered tailings (Venu Babu, 2007).

Ecological dereliction occurs in mine areas due to interference in the plant, soil, geological, hydrological and social fabric of the ecosystem. Minerals are formed over a period of centuries through the process of fossilization of biological materials and their conservation into different mineral strata. During the process of mining, these natural resources are removed from the body of the earth, processed and used in industries to keep pace with the developmental cycle. Ecological risk assessment due to deleterious impacts of uranium mining and milling process require development of ecologically viable, sustainable radioactivity consolidation in the tailings and the overburden dumped during mining. Open pit mining drastically disturbs the existing flora and fauna as well as the soil's biological, physical and chemical systems. The ultimate objective of radioactive waste management is to protect individuals, society and the environment from the harmful effects of ionizing radiation due to spent fuel and radioactive waste, both at present and in the future. It is therefore expected that the process of industrialization, including mining, may lead to ecological disturbance if control measures are not exercised. However, the degree of disturbance depends on the mode of mining, that is, underground or surface, the location of the mining site and the ecological, hydrological, geological and sociological conditions of the mined areas.

Since mining and processing of economically viable mineral deposits are prerequisites of industrialization, emphasis has to be given on developing ecological strategies for environmental restoration and consolidation of radioactivity. Ecological risk assessment and management of land consolidation are the two objective requirements for the sustainable development of mined and tailings areas. It is important to monitor and control ecological risks arising due to mineral excavation and processing. On the basis of systematic elaboration of the connotation and characteristics of ecological risks of the land consolidation, analysis needs to be done for ecological risk factors and evaluation index system relative to the ecological effect of land consolidation, and thereof identification, assessment and control of ecological risks based on full-process analysis of land consolidation projects, including ecosystem service

DOI: 10.1201/9781003326151-2

functions as an important content of ecological risk assessment. Thus, the limitation of research in the past on ecological risk has been overcome with more emphases on changes in the structure and functions of an ecosystem. Scientific issues and development trends that need more attention in tackling ecological risks of land consolidation are pinpointed (Liu et al., 2017). Taking an ecosystem approach to restore the productivity potential of disturbed mined land provides an answer for sustainable restoration and ecological strategy to consolidate radioactivity below ground.

2.1 IDENTIFYING THE NATURE AND SOURCE OF RADIOACTIVE POLLUTANTS IN URANIUM MILL TAILINGS

Because uranium occurs in soils and fertilizers, the element enters food and human tissues. Like uranium, radium is also present in all rock and soils in variable amounts, and elevated levels can be expected in phosphate rocks. As with uranium mines everywhere, the principal issues are radioactivity, dispersion and water contamination. Milling is the first step in making natural uranium ore into fuel for nuclear reactors. Uranium mills use chemicals to extract uranium and make "yellowcake", a powder that can be processed into fuel. They produce "mill tailings", a sandy waste containing heavy metals and radium, which is radioactive (USNRC, 2020). Uranium mill tailings contain the radioactive element radium, which decays to produce the radioactive gas radon. Most uranium mill tailings are placed near the processing facility, or mill, where they come from. While uranium itself is only slightly radioactive, radon, a radioactive inert gas, is released to the atmosphere in very small quantities when the ore is mined and crushed. Radon is one of the decay products of uranium and radium, and occurs naturally in most rocks—minute traces of it are present in the air which we all breathe.

Production of uranium, like all other minerals, includes exploration, excavation/mining and then milling operations. During mining, raw uranium ore is extracted, brought to the surface and crushed into a fine sand. Uranium tailings are generated as a waste by-product of uranium mining. The valuable uranium-bearing minerals are then removed via heap leaching using acids or bases, and the remaining radio active sludge, called "uranium tailings", is stored in huge impoundments. Uranium recovery from the ore involves a number of steps, including changing the physicochemical characteristics of the bulk ore and oxidation, leaching, ion exchange and precipitation as Magnesium-Di-Uranate (MDU). Though the major portion of uranium present in the ore is extracted, a fraction, though quantitatively small, remains unextracted and is finally discharged with the tailings. The unextracted fraction may be either soluble hexavalent complexes of uranium or insoluble tetravalent form. Migration of this unextracted fraction of uranium from the tailing pile to the plants growing over the soil-covered tailing pond is a matter of concern. A short ton (907 kg) of ore yields one to five pounds (0.45 to 2.3 kg) of uranium depending on the uranium content of the mineral (Grammer, 1981).Uranium tailings can retain up to 85% of the ore's original radioactivity (Robinson et al., 1979).

The uranium and radium content in drinking water sources is outlined in WHO (1996) draft documents. Accordingly, the guidelines for drinking water are also given in this document. The guideline suggested by WHO (1993) are 1000 Bqm^{-3}for ^{226}Ra and 140 ug.l^{-1} for ^{238}U. However, these values are far greater than the limits followed in India (300 Bqm^{-3} for ^{226}Ra and 60 ug.l^{-1} for natural uranium).

Regardless of how uranium is extracted from rock, the processes leave behind radioactive waste. For example, the solid radioactive wastes that are left over from the milling processes known as tailings and the liquid wastes are called raffinates. Mill tailings and raffinates are stored in specially designed ponds called impoundments. The tailings remain radioactive and contain hazardous chemicals from the recovery process. Uranium mill tailings are of particular environmental concern because (a) they retain much of the radioactivity of the ore from which they were derived; (b) their radioactivity is very long lived; (c) they contain a range of biotoxic heavy metals and other compounds; (d) they may contain sulphide minerals, which can generate acid mine drainage; (e) their granular to slime consistency makes them susceptible to leaching, erosion or collapse under various conditions; (f) the common method of surface disposal exposes a large surface area to the natural elements and thus increases the risk of radiation flux release, radioactive and geochemically toxic dusts, and interaction with surface water systems; and (g) the large surface area of these generally thin tailing deposits (or "piles") adversely affects large areas of land and renders potentially valuable land unfit for other uses (IAEA, 2004b).

In general, radioactive waste is separated into three categories: low-level waste (LLW), intermediate-level waste (ILW) and high-level waste (HLW), depending on its level of radioactivity and the length of time it remains hazardous. Disposal of LLW and most ILW is a mature practice, while most HLW is safely stored in dedicated facilities. The permanent disposal of HLW in deep geological repositories is accepted to be practicable by the scientific and technical community, but has yet to be accepted by civil society in many countries.

To appropriately manage radioactive waste it is required to

- technically work on reducing the amount of waste generated, and
- safely dispose of radioactive waste so as to minimize the impact on the environment.

Waste materials from mining activities that are discharged as mill tailings and left in surroundings without any countermeasure can generate various ecological, social and hydrological impacts. In the past, tailing areas were permeable, allowing seepage into surrounding groundwater. Today, they are constructed to be impermeable through the use of bentonite clays, compacted till or synthetic fabric (geomembranes) dams built to retain the tailings. Research toward the development of new geomembranes and new technologies to safely contain waste material from mining activities is ongoing. Large quantities of water and various chemicals are used in the milling processes, and tailing waters must be treated and neutralized before they can be discharged to natural drainage basins.

Uranium tailings contain over a dozen radioactive nuclides, which are the primary hazard posed by the tailings. The most important of these are thorium-230, radium-226, radon-222 (radon gas) and the daughter isotopes of radon decay, including polonium-210. All of these are naturally occurring radioactive materials, or

"NORM". Emanation of radon from the soil or rocks depends on factors such as 226Ra content, porosity, atmospheric pressure, temperature and moisture content. 226Ra in ore, waste rocks and tailings is computed from the ore grade or measured by gamma counting or chemical analysis. 226Ra in the soil is measured by gamma counting or by chemical analysis.226Ra in the soil of Singhbhum Thrust Belt (STB) varies from 5 to 285 Bq kg21, averaging around 74 Bq kg21 (Khan and Puranik, 2011).

2.2 PATHWAYS BY WHICH THE POLLUTANTS MAY REACH HUMANS

Though the major portion of uranium present in the ore is extracted, a fraction remains unextracted and is finally discharged with the tailings. The unextracted fraction may be either soluble hexavalent complexes of uranium or insoluble tetravalent form. After the ore is processed, the tailings that remain are released into tailing ponds specially designed for this purpose. This radioactivity in the tailings might emanate into the atmosphere or enter the food chain.

Although uranium has long been recognized as a danger for living organisms, very little was known about the physical and chemical pathways uranium takes to enter the food web. When then secretary of the US Department of Interior Ken Salazar declared a 20-year moratorium for new mining claims on federal lands adjacent to Grand Canyon National Park in 2012, he turned to the USGS to address uncertainties in the ecological impact of uranium mining (USGS, 2016). The most critical food chain in the world for concentrating airborne radionuclides is the lichen-caribou-human food chain. Lichens accumulate atmospheric radionuclides more efficiently than other vegetation due to their lack of roots, large surface area and longevity. Uptake from the substrate is minimal compared with the uptake from wet or dry deposition (Svoboda and Taylor, 1979 and Boileau et al., 1982). The risk of fatal cancer from a dose of 1.7 mSv is 8.5×10^{-5} per year, and 6×10^{-3} over a 70-year lifetime if caribou meat, liver and kidney are consumed at the rate postulated. Doses and risk are almost ten times higher in other regions of Canada where 210Po concentrations in caribou are higher and if higher intake rates are assumed (Thomas and Gates, 1999).

Radioactive substances present in the earth are the sources of radioactivity in the atmosphere. Radioactive pollutants include radon-222, iodine-131, strontium-90 and plutonium-239. The contamination of the environment with toxic metals has become a worldwide problem, affecting crop yields, soil biomass and fertility and contributing to bioaccumulation in the food chain. The legacies of past uranium mining and milling activities continue to be a cause of concern and require assessment and remedial action. This problem has been recognized in many parts of the world over the past five decades. If uranium tailings are stored above ground and allowed to dry out, the radioactive sand can be carried great distances by the wind, entering the food chain and bodies of water (Anon, 2022) The danger posed by such sand dispersal is uncertain at best given the dilution effect of dispersal. The majority of tailing mass will be inert rock, just as it was in the raw ore before the extraction of the uranium, but physically altered, ground up, mixed with large amounts of water and exposed to atmospheric oxygen it can substantially alter the chemical behavior. An

EPA estimate of risk based on uranium tailing deposits existing in the United States in 1983 gave the figure of 500 lung cancer deaths per century if no counter measures are taken Diehl (2011).

Pollutants may enter the human body by either inhalation, ingestion or absorption through skin. According to the IAEA Fundamental Safety Principles, the objective of radioactive waste management is to deal with radioactive waste in a manner that protects human health and the environment now and in the future without imposing undue burdens on future generations (IAEA, 2009).

Radionuclides contained in uranium tailings emit 20 to 100 times as much gamma radiation as natural background levels on deposit surfaces. The radium-226 in tailings continuously decays to the radioactive gasradon-222, the decay products of which can cause lung cancer. Some of this radon escapes from the interior of the pile. Radon releases are a major hazard that continues after uranium mines are shut down. The EPA estimates the lifetime excess lung cancer risk of residents living near a bare tailing pile of 80 hectares at two cases per hundred. Since radon spreads quickly with the wind, many people receive small additional radiation doses. Although the excess risk for the individual is small, it cannot be neglected due to the large number of people concerned. The EPA estimates that the uranium tailing deposits existing in the United States in 1983 would cause 500 lung cancer deaths per century, if no counter measures are taken.

Radon gas may enter the human body through air breathed in the vicinity of tailings. Long-lived natural radioisotopes—that is, all nuclides of the uranium decay series—are found in concentrated amounts in uranium mill tailings. One of the major pathways of radiological exposure to humans from uranium milling operations is through the beef/milk food chain (Rayno, 1983).

In many cases, exposure may occur simultaneously from many sources and through multiple routes. Pathways of exposures to lead, for example, include air pollution from traffic and industrial emissions, drinking water, food, tobacco smoking, dusts, paints and other industrially produced commodities and soil. Valid exposure assessment therefore typically requires detailed knowledge about the geographical distribution of the pollutants of concern, the temporal variations in pollution levels and the processes of exposure. People are often exposed to different pollutants simultaneously. Exposure to these may occur at different locations (e.g., in the workplace and/or at home) and at different times. Radionuclides affect the living organisms both as heavy metals and by their radiation. They participate in radionuclide and heavy metal transfer to the biosphere and humans as elements of the food chain of marine biota.

In researching radiological exposure to humans through the beef/milk food chain, various investigators have shown the extent of uptake and distribution of 238U, 234U, 230Th, 226Ra, 210Pb and 210Po in plants and cattle. These long-lived natural radioisotopes, all nuclides of the uranium decay series, are found in concentrated amounts in uranium mill tailings. In this paper, data from these investigations are used to estimate the dose to humans from consumption of beef and milk from cattle that have fed on forage contaminated with the tailings. The estimated doses from this technologically enhanced source are compared with those resulting from average dietary intake of these radionuclides from natural sources (Rayno, 1983).

To evaluate the radiological risk to human health associated with the ingestion of radionuclides in foods, some different plant foods cultivated in the mining area were collected and analyzed. Results showed that local foods contained higher [210]Pb (8.1–150 Bq/kg) and [228]Ra (0.3–4.2 Bq/kg) than the worldwide reference values. The annual committed effective doses to children (3.1 mSv/y) and adult (1.3 mSv/y) from the ingestion of local foods might exceed the recommended reference level of 1 mSv/y (Yang et al., 2021).

To assess radiation doses from radon short-lived progeny by the inhalation of exhaust fumes by mechanic workers exposed, concentrations of these radionuclides were measured in petrol (gasoline) and gas-oil exhaust fumes by evaluating mean critical angles of etching of the CR-39 and LR-115 type II SSNTDs for alpha particles emitted by the radon and thoron decay series. Committed effective doses due to [218]Po and [214]Po short-lived radon decay products from the inhalation of petrol and gas-oil exhaust fumes by workers were evaluated. A maximum value of 1.35 mSv y^{-1} due to radon short-lived decay products from the inhalation of gas-oil exhaust fumes by mechanic workers was found, which is lower than the (3–10 mSv y^{-1}) dose limit interval for workers (Misdaq et al., 2015a, b). Raised levels of radiation caused by a chemical in car exhaust fumes have been found in the teeth of children who live near motorways.

The levels are low and come from polonium-210—which is also emitted from power station chimneys, volcanoes and the ground. Henshaw (1991), an expert on radon in the environment, found significantly raised levels of polonium up to 10 km from motorways. He says in *The Lancet* that as a result of the findings, radioactivity should be added to the list of potential cancer-causing agents in vehicle exhaust. The researchers knew radiation in children's teeth was higher in urban areas. They say the polonium-210 (210Po) in exhaust can come from leaded petrol or as a decay product of oil. The research into levels of polonium, which emits alpha radiation, involved the analysis of 2000 milk and permanent teeth from children. "The results are surprising in view of the number of potentially confounding factors such as the closeness of major urban areas and the fact that we do not know the length of time children have lived at their present addresses".

The evidence for possible links between domestic radon exposure and incidence of leukemia and other non-lung cancers is reviewed. Recent calculations of the radon-derived dose to red bone marrow suggests that if background radiation is linked to leukemia in the general population, then radon exposure may be a causative factor (Henshaw et al., 1990). According to Eatough and Henshaw (1993), there is a potential for causing cancer based on studies of high exposure to radon gas, another emitter of alpha radiation. This is associated with leukemia, brain tumors and kidney cancer, especially in children.

The International Agency for Research on Cancer (IARC, 2022) classifies types of air pollutants based on their link with cancer. For example, gasoline engine exhaust is classified as a possible carcinogen. Radon, diesel engine exhaust and outdoor air pollution are classified as known carcinogens. Researchers have also found that radon gas in indoor air can lead to lung cancer. The EPA reported that radon is the second leading cause of lung cancer in the United States, killing 21,100 people per year. The EPA relies on the BEIR VI models, based on an evaluation of radon

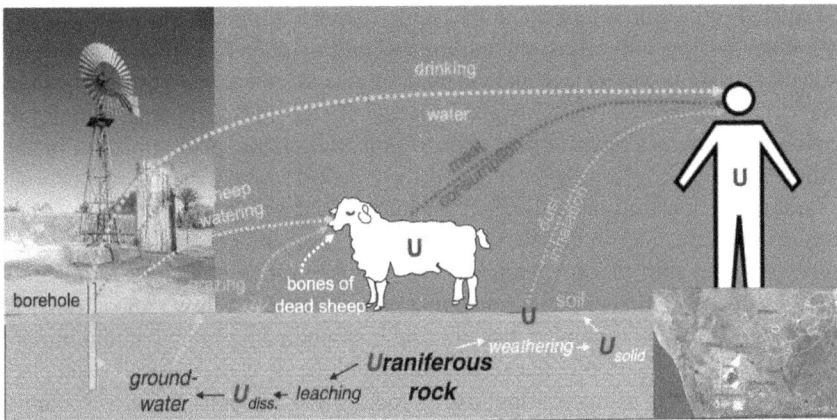

FIGURE 2.1 Graphical abstract of exposure pathway for farm residents resulting in U-uptake (after Winde et al., 2017).

exposure and lung cancer risk in studies of miners. But these models did not account for co-exposure to diesel exhaust, a known human carcinogen recently classified by IARC. It is probable then that a portion of the lung cancer deaths in the miner cohorts are originally attributable to the exposure to diesel rather than radon (Xiaodong et al., 2017) Long-term exposure to radon is considered to be the second most frequent cause of lung cancer after cigarette smoking (WHO, 2009). Similar to radon, a recent study showed that the unit risk estimate for diesel exhaust is strikingly high at 2.1 excess lung cancer deaths per 1000 individuals for a lifetime exposure at an environmental concentration of 0.8 $\mu g/m^3$ (Vermeulen et al., 2014).

According to Winde et al. (2017), contaminated drinking water is the main exposure pathway for farm residents resulting in U-uptake rates exceeding the WHO's tolerable daily intake (TDI) limit by up to 900%. This is somewhat mitigated by the fact that U-speciation is dominated by a neutral calcium-uranyl-carbonate complex of relatively low toxicity (Figure 2.1).

The decay products of radon gas are a well-recognized cause of lung cancer in miners working underground. When radon was found to be a ubiquitous indoor air pollutant, however, it raised a more widespread alarm for public health.

2.3 SITE SELECTION AND DESIGN OPTIONS THAT MAY BE IMPLEMENTED TO FACILITATE DISPOSAL AND/OR CONTROL THE EXTENT OF RELEASES FROM THE IMPOUNDMENT

Uranium mill tailings are the solid residues and the associated liquids remaining after uranium has been extracted from an ore. The solids consist primarily of the finely ground bulk material of the original ore and also contain a variety of chemical substances that are precipitated from the tailing liquids. The tailing solids and liquids contain both radioactive and nonradioactive materials which, if released and dispersed in the environment in unacceptably large amounts, may have some detrimental impact on humans and their environment. Although the radioactivity of the tailings is of natural origin, the mining and milling process results in (1) bringing the radioactivity to the surface and (2) changing its chemical and physical form in a manner that increases its mobility and consequently enhances the possibility for dispersion in the human environment and the potential radiological impact in humans (IAEA, 1981).

Tailings produced in the milling of uranium ores differ from those of most other metal milling practices in that they have a significant content of radioactive components. Uranium mill tailings management therefore presents some additional and particular considerations for environmental protection due to their radioactivity. The basic design and operating objectives for the impoundment of the uranium mill tailings in all cases is (1) to provide an impoundment that is both physically and chemically stable and (2) to control the loss of radioactive elements to the environment to acceptable levels. The important considerations that influence the effectiveness of tailings impoundment relate to both site selection and choice of the impoundment system to be used. Each potential tailings impoundment site has inherent site-specific advantages and disadvantages. Site selection should make use of the site-specific advantages. The design and engineering should then account for the site-specific disadvantages. Site selection and design and engineering of the tailings impoundment should be considered in conjunction with the overall mining and milling operations. Processes and methods used in both the mine and mill will have a direct influence on the requirements, performance and cost of the impoundment. The concepts of storage and disposal should be considered when evaluating tailings management systems. Technology is available to ensure that any properly sited and designed tailings impoundment can meet the requirements of storage, that is, confinement of wastes over a limited period of time with the possibility for retrieval and eventual disposal of tailings. However, tailings disposal is the desired management objective. In this case there is no intent to retrieve and, more importantly, no reliance on continued surveillance and maintenance to ensure integrity of the impoundment system. The time at which survey can be discontinued after decommissioning the mill and rehabilitating the tailings impoundment could vary considerably with design and, in particular, with certain site-specific parameters. National differences exist regarding the objectives of disposal as it relates to tailings; however, the primary basic difference relates only to the acceptance, within the objectives of disposal, of long-term restriction of access to, or use of, the rehabilitated tailings area (IAEA, 1981). One of the most important aspects to be addressed in tailings management is the selection of a site that is both geologically and geomorphologically stable. The site-selection process consists of collecting appropriate information regarding the natural conditions, geology and geomorphology of the region concerned, and then choosing a site that has a favorable combination of conditions with respect to those factors governing the long-term integrity of the impoundment.

The tailings, or solid waste remaining after uranium has been extracted from ore, and the associated liquid wastes must be disposed of in an economical and environmentally acceptable way. Selecting an impoundment site is the first task for establishing an acceptable disposal facility. Because of the long-term nature of the pollution and the environmental impact potential of uranium tailings, the selected site and engineered impoundment must serve as an effective containment facility not only in the short term but also in the very long term. Such long-term stability is an extremely difficult task to engineer into a facility that is subject to slow but perpetual geologic processes such as wind and water erosion, chemical and physical weathering, seepage and leaching. Neutral features that occur on or surrounding some sites provide a natural "screen" or "barrier" to such forces or processes in a manner that cannot be achieved through engineering or construction (Robertson, 1980).

A five-step process was used to locate and identify a suitable alternate disposal site: "(1) designation of a search region; (2) development of guidelines for eliminating unacceptable areas from the search region; (3) application of the guidelines; (4) evaluation and field reconnaissance of potential sites; and (5) selection of a single disposal site for comparison with the proposed action, stabilization in place" (DOE, 1986a). A list of 22 screening guidelines was developed and used to eliminate broad areas that might have required greater complexity in design or that posed regulatory problems. Three candidate sites were identified in the screening step. These sites were then further evaluated on the basis of hydrologic, meteorologic, geologic, environmental and economic data, as well as field reconnaissance. A preferred site was identified in five mile Wash, approximately 16 miles from the former processing site. The Five mile Wash site was selected over the other two candidate sites primarily on the basis of groundwater conditions (DOE, 1986a, 1986b, 1986c).

To reclaim a uranium mill tailing pile according to principles of safe long-term isolation, detailed investigations have to be performed in advance to assess the site.

If the tailing pile presents an immediate hazard, then intermediate protective measures can be taken in parallel, such as installation of a cover against windblown dust or collection of seepage waters. These measures, however, should not conflict with long-term measures to be taken later.

According to Diehl (2011), the site must be appropriate for tailings disposal from the view of geology and hydrology:

- it should not be located on a geologic fault,
- it should not be threatened by the risk of earthquakes,
- natural impermeable layers should be present,
- the site should not be located in the flood plain of rivers,
- the phreatic level should be rather deep,
- any seepage should not present a risk to groundwater,
- deposits of clay materials appropriate for lining and covering the deposit should not be located too far away,
- the site should be remote from residential areas, and so on.

In Jaduguda (India), the tailing ponds are selected at a site where the geological structure is almost unaffected by any hydrogeological processes. However, the tailing pond might contaminate the groundwater regime by continuous seepage and leaching of radionuclides and other toxic metals due to the interaction of rainwater in the tailing ponds (Sarangi et al., 2009).

3 Physicochemical Analysis of Soil and Tailings

The physical properties of soil have a direct relationship with plant growth and establishment because the depth of rooting zone and the air and water relations within it are determined largely by the physical makeup of soil. Furthermore, they have additional indirect significance because many chemical and biological aspects of soil fertility can be inferred from physical properties. The key point is that all the physical properties affect the management practices required to maintain moisture potential, porosity, temperature and mechanical resistance in a range suitable for soil development and high production on mine lands (Munshower, 1993; Soni et al., 1992).

The rapid changes in physical properties may have been caused by the accumulation of organic carbon with the age of biorestoration. This organic carbon might have undergone a rapid process of humification with its end products being capable of reacting with the inorganic constituents and thus contributing to the development of soil (Varela et al., 1993). In the course of this process, the plant roots also may have created channels and macropores by breaking the soil mass (Goss, 1987). Organic substances bind simple particles and aggregates leading to a porous structure, and levering of the bulk density improves drainage and aeration conditions (Oades, 1984).

The physical and chemical nature of uranium mine tailings and effluent is considerably different from ordinary soil materials and waters. Therefore, many of the conventional methods for the collection and analysis of natural soils and waters are not directly applicable to tailings. Secondly, there is a wide range in the physical and chemical nature of uranium tailings. The composition of the ore, the milling process, the nature of tailing deposition and effluent treatment vary considerably and are highly site specific. After recovering the economically viable portion of the ore, the remaining solid slurry or tailings from uranium ore processing is discharged into an engineered system called a tailing pond. Among the radionuclides, quantitative content of residual uranium is highest in the tailing pond, and various environmental interactions such as precipitation, changes in pH, redox potential, microbial activities and organic associations have the potential to fix (precipitate) or solubilize it. The chemical fractionation of "U" in mill tailings of both operational and nonoperational tailing ponds of Jaduguda uranium mining and ore processing site is part of the present study. Also, the role of various physicochemical parameters (pH, electrical conductivity, Total Carbon (TC), etc.) on the mobility of uranium has been investigated (Singh et al., 2018).

The important components of the physical and chemical environment of the tailings impoundment include:

- physical, chemical, mineralogical and radiological characteristics of the tailings solids;

DOI: 10.1201/9781003326151-3

- chemical and radiological characteristics of the tailings porewater;
- physical, chemical and radiological characteristics of surface water flow from, and seepage out of, the tailings impoundment;
- hydrogeological conditions within and around the tailings impoundment and the characteristics of exfiltration out of the tailings;
- air quality and meteorological conditions over and around the tailings impoundment.

The tailing solids and tailing porewater are the sources for potential chemical and radiological contamination (NUTO, 1985).

3.1 PHYSICAL AND CHEMICAL PROPERTIES OF SOIL AND TAILINGS

The physical properties of the soil are very important for vegetative growth and the sustainable use of soil, while the chemical properties of the soil are important to soil fertility, plant growth and reproduction. That is why it is important to understand how chemical properties of the soil interact to affect the soil's capacity to store and release nutrients, and how soil chemistry can affect soil structure.

Soil carbon and nutrient availability play crucial roles in ecosystem sustainability, and they are controlled by the interaction of climatic, biotic, and soil physicochemical variables. Although soil physicochemical properties have been recognized as vital variables for predicting soil organic carbon (SOC) and nutrients, their relative influence across broad geographical scales has yet to be evaluated when simultaneously considering many other drivers (Li et al., 2020).

The amount and rate of water, oxygen and nutrient absorption by plants depend on the ability of the roots to absorb the soil solution as well as the ability of the soil to supply it to the roots. The spatial information about land resources is needed to provide a direction for the planning and implementation of restoration of post-tin-mining land so that the environmental degradation can be mitigated and land productivity can be restored. In addition, appropriate plant species and location-specific management techniques (i.e., soil fertility improvement and application of soil conservation techniques) should be determined and developed. Determination of plant species and location-specific technology is based on the characterization and evaluation of potential land resources to support the planning and implementation of restoration of post-mining land so that the utilization of post-mining land can be optimal and the environmental functions can be restored. Soil physical and chemical properties affect construction, plant growth and functional development of habitat in created marshes (Callaway et al., 2012).

Soil physical properties are largely derived from parent material (e.g., weathered bedrock, aeolian silt, marine or lacustrine sediment, etc.) and other geomorphic attributes such as slope, elevation and aspect. These factors ultimately interact with climate, weathering and time to give each soil its own distinctive combination of sand, silt and clay-sized particles, which may or may not include larger fragments of rock; in sum, the soil's texture. The importance of soil texture is discussed in more detail in Stanturf et al. (2021). From a soil ecology standpoint, texture largely determines the water-holding capacity of a particular soil, and the amount of water is strongly related to the types and numbers of soil organisms that will inhabit and influence that soil. In general, sandy soils retain less water, clayey soils retain more, and silty soils occupy the middle ground. Inasmuch as water availability influences plant productivity, these different textured soils then also have general characteristics of Soil Organic Matter (SOM) storage, with sandy soils typically being relatively low on the SOM spectrum and clayey soils on the high end.

Sharp boundaries within soil profiles, marking abrupt changes in soil physical and chemical properties, frequently originate from disconformities rather than exclusively from pedogenic processes. In soils and weathering profiles, simple stratigraphic principles do rarely hold, as discussed in detail by Pain and Ollier (1995). Likewise, as discussed here, top-down pedogenesis operating in purely residual materials cannot be assumed either. Therefore, the interpretation of the vertical distribution of soil properties requires a concept of regolith and soil genesis, that is, an allochthonistic approach. As a crucial part of this approach, the original properties of the parent material and its pedo genetically formed alteration products have to be reconstructed (Blume and Röper, 1977).

Factors such as soil characteristics, climatic conditions, type of plants, the part of the plant concerned, the physicochemical form of the radionuclides and the effect of competitive species can influence transfer factor (TF) values. Assessment models normally make use of a plant/substrate concentration ratio, referred to as a TF to estimate the transport of radionuclides and other elements of interest through the food (Vera et al., 2003 and Sheppard and Sheppard, 1985). Adewole and Adesina (2011) highlighted that inaddition to enriching the soil with heavy metals, mining activities and mine waste generation can disrupt nutrient dynamics in soils as a result of dynamic and interaction alterations in physical, chemical and microbiological processes. An understanding of the soil/substrate physicochemical characteristics around mines may be important to inform remediation strategies aimed at reducing Heavy Metal (HM) concentrations or bioavailability (Fashola et al., 2016).

Indorante et al. (1981) also found that after reconstruction the properties of newly constructed mine soils reflected premining overburden character and the method of soil construction. Generally after a period of two years, the revegetated soils, having less compaction, are also somewhat more favorable to root penetration and water circulation; at least in the upper 25cm of the soil studied. Potter et al. (1988) has also reported a similar evolution of mine soils in North Dakota, USA.

Criteria used to evaluate the sustainability of a vegetation cover include diversity of species, dynamic stability and productivity. The productivity of plants depends largely on climate and the fertility of the growing medium. Tailings fertility is related to (1) the physical properties of the tailings, which influence the amount of water that can be held in the profile and the ease with which roots can penetrate, and (2) the tailings chemistry (pH, EC and nutrients). The dynamic nature of revegetated systems requires data not only on the initial conditions but also over time to monitor the system's evolution.

Morten et al. (2002) has given a detailed discussion on the dependence of plant organ concentration and concentration ratio on soil parameters (pH, delta pH, total soil concentration, etc.). Further, the interrelationship between tissue concentrations (stem and leaf) was also discussed by the author. Two horizons of soil, that is AE and Bs, were selected for evaluation of concentration ratio; apart from this, the long-term effect of plant uptake on recycling was also presented in the review. These authors studied the extent of uranium and thorium uptake and recycling by blueberry (*Vassinium pallidum+* Aiton) in native habitat and identification of the soil properties and processes that contribute most to uranium and thorium bioavailability in this system. Both AE and Bs horizon soil concentrations were determined and the concentration ratio for stems and leaves were evaluated based on the soil concentration of both horizon and tissue (stems and leaves) concentrations. Mean uranium concentration was 8.65×10^{-3} $\mu g.g^{-1}$dry in leaf and 7.95×10^{-3} $\mu g.g^{-1}$dry in stem. Uranium concentration in leaves shows little preferential translocation. Moreover, uranium uptake was found to be sensitive with respect to soil properties. Plant tissue concentrations of radionuclides have rarely shown a linear relationship to radionuclide concentration in substrate (Sheppard and Sheppard, 1985,and Sheppard et al., 1992). Plant and soil concentrations were not significantly correlated; the correlation coefficient between AE horizon soil and leaf concentration was −0.13, whereas it was −0.41for stem. Plots of Concentration ratios identify a clear differentiation by tissue type and by nuclide. Although CRs based on AE horizons are somewhat higher, CR trends were virtually identical for both AE and Bs horizons. CRU_{lv} values were not significantly different from CRUst. When CRs were calculated based on AE horizons, CRTh in both stem and leaves declined with increase in soil concentration. Additionally, $CRTh_{AE}lv$ and $CRTh_{AE}st$ were highly correlated (R = 0.88, p = 0.001), but $CRU_{AE}lv$ values were poorly related to $CRU_{AE}st$. Studies by Sheppard and Evenden (1988) found that U and Th CR values decrease as soil concentration increases. Control pot study of blueberry plants grown in organic soil artificially contaminated with uranium were showing CR values nearly an order of magnitude higher as reported by Morten et al. (2002), although CR values decline significantly with time. This suggests plant uptake may eventually equilibrate. Neither uranium plant concentration nor uranium CRs were linearly related to uranium soil concentrations, suggesting that soil uranium may be partitioned more explicitly into available and unavailable forms. Weathering processes responsible for particle size reduction and increased surface area appear to enhance the release and availability of uranium. Studies carried out at the Anaconda mill site for forage uptake of uranium series radionuclides has given an uptake coefficient for ^{238}U (ratio of vegetation activity to soil activity) that varies from 0.69 in Indian rice grass roots to 0.01 in four-wing saltbush (*Atriplex canescans*) shoots (Rayno et al., 1980).

Mining activities and mine waste generation also can disrupt nutrient dynamics in soils as a result of dynamic and interaction alterations in physical, chemical and microbiological processes (Adewole and Adesina, 2011). Numerous studies have emphasized the significance of soil properties including organic matter (OM), particle size distribution including clay content, redox potential, electrical conductivity (EC), moisture content, cation-exchange capacity (CEC) and pH on the behavior of HM in soils (Šourková et al., 2005; Keskin and Makineci, 2009).

These results suggest that the establishment of native plants has successfully favored this property for these soils that have been transported from different areas for capping the tailings. Similar improvement in bulk density by afforestation has been reported by Parthiban and Rai (1992) and Singh and Raman (1979). Thus, it may be assumed that with the passage of time and concomitant addition of organic carbon, the mine soils under the canopy of native species might show a considerable retreat in bulk density.

The amount of pore space available is used as a factor to evaluate the suitability of soil as a medium for plant roots (Groenevelt et al., 1984). Sandy soils usually have less pore space whereas clay loam sand clay shave high total pore space (Troeh and Thompson, 1993). Generally, practices such as revegetation that improves soil structure also helps to modify porosity of soils (Patnaik, 1978). In undisturbed forest, the macropores dominate the pore space and facilitate rapid movement of water through soil profile (Humbel, 1975). Conversely, the limited porosity of mine soils adversely affects aeration and water retention (Bussler et al., 1979).

Schafer et al. (1979) have also reported similar evolution of 1- to 50-year-old strip mine soils with high compaction. The most universal effect of soil pH on plant growth is nutritional. The soil influences rate of plant nutrient relies, solubility and availability of plant nutrients depends on **porosity** because if the pore size is more water will wash away all the nutrients to water table as gravitational water and if porosity is less there will not be any air, respiration of roots and hence, absorption of nutrients will suffer. (Troeh and Thompson, 1993). A slightly acid to slightly alkaline pH provides optimum growth conditions, because toxicity is less and nutrient availability is greater within this range (Hausenbuiller, 1981). The soil reaction is therefore a good guide for predicting which plant nutrients are likely to be deficient.

Kirlew and Bouldin (1987) reported a decrease in pH by 0.11 and 0.16 units around the roots of two corn hybrids. The decrease in pH may be attributed to an increase in organic matter and the leaching of salts, particularly calcium. This decrease in pH towards equilibrium through the leaching of calcium suggests that the native plants established on mine soils are acid ophilic basilo tolrents.

Ingested (1979) has also found a reduction in pH under the canopy of Scots pine and Colorado spruce on soils rich in exchangeable bases and often calcareous in nature. Troeh and Thompson (1993) reported that the addition of decomposable organic matter to a soil results in the formation of organic acids that enhance the cation exchange capacity and reduce pH. They further noted the root uptake of calcium and other cations combined with release of H^+ ions from the roots is likely to reduce pH in the rhizosphere by 0.1 or 0.2 units as compared to rest of soil. Jenny (1980) found that *Populus* species reduce the base saturation and pH of the soils and increase leaching of soils $CaCO_3$.

Hass et al. (1998) studied bioaccumulation of uranium by lichens. The uptake of aqueous uranium by *Peltigera membranacea* was found to be a function of time and pH. It was determined that uranium sorption was strongest in the pH range 4–5 with maximum sorption occurring at pH of 4.5 and an incubation time of 24 hours. Similarly, experiment and field studies by Mann and Fyfe (1995), Milodowski et al. (1990), Lovey et al. (1991) and Macaskie et al. (1992) demonstrated that algae and bacteria can facilitate the precipitation of solid U phases directly from solution, through absorption of aqueous uranium into bacteria cell walls and subsequent nucleation of hydrous uranyl phases (Macaskie et al., 1992) or through U absorption followed by reduction of UO^{2+} torelatively insoluble U^{4+} at the cell wall interface (Lovey et al., 1991). Algae may bioaccumulate uranium through a combination of mechanisms including surface complexation, biomineralization and physical trapping of dust and soil particles.

Mine soils on the Great Plains reveal that topsoil electrical conductivity values within the range of 4–8 ds/m (Soil Survey Staff, 1951). Conductivities that fall in this range are not inhibitory to plant grow thin most semiarid to arid regions unless other factors such as particle size and sodium content compound the salinity problem (Munshower, 1993).

surface with age. The reduction in soluble salts and pH recorded with age of bioreclamation further lends supports to this assumption. These findings concur with the observations made on variety of other mine soils (Schafer et al.,1980; Ciolkosz et al., 1985). Despite significant improvement with age, the electrical conductivity of revegetated sites was less than the adjacent reference soils. This difference in electrical conductivity between the two sites can be attributed to low water percolation due to low porosity and therefore slow leaching of salts from upper layers. These results on electrical conductivity are comparable to those reported by Brynes and Miller (1973); Hauser and Chichester (1989); Chicchester and Hauser (1991).

3.1.1 TOTAL NITROGEN PERCENTAGE

The potential rate of nitrogen uses by growing plants gradually exceeds the rate at which nitrogen is available. Consequently, the amount of available nitrogen in the soil is usually very small. According to Troeh and Thompson (1993), as oil may contain $110 kgha^{-1}$ or more of available nitrogen; the amount of organic nitrogen being much larger can be considered are servoir of which between 1% to 5% is likely to become available each year in temperate climate sand upto 50% under tropical conditions. Skeffington and Bradshaw (1981) felt that a large pool of organic nitrogen and a high rate of ammonification was necessary to sustain vegetation and to prevent nitrogen immobilization. Detailed laboratory and field investigations have shown that the critical factor affecting the success of revegetation is the development of the nitrogen cycle (Bradshaw et al., 1975; Stevenson, 1965). Woodmansee et al. (1980) estimated that it might take up to 2160 years of natural succession on over burden for the nitrogen pool to grow large enough to support as table and self-sustaining plant-soil system.

According to Troeh and Thompson (1993), the decomposition reduces the amount of organic carbon but leaves it enriched in certain nutrients, especially nitrogen. These results are in accordance with those reported by Jones and Wild (1975).There duction in pH has also been evidenced to increase the availability of nitrogen in soils. Mengel and Kirkby (1979) found that the correction of soil reaction often enhances the activity of soil microbial biomass and subsequent decomposition of organic matter and there by increases the nitrogen content and its availability. Some workers have attributed the increase in total nitrogen to weathering of parent material (Cornwell and Stone, 1968; Varela et al., 1993).

These results probably indicate that the rate of nitrogen return to the soil through litter fall and its subsequent decomposition is greater than the rate of nitrogen uptake by vegetation on undisturbed soils where the nutrient loss through leaching may also be less. The significant differences between the soils of reconstructed and undisturbed sites interms of their nitrogen content might be related to their differences in texture and organic carbon content, which are important in determining the rate of nitrification. Thus, the undisturbed soils with adequate organic carbon would have a higher rate of nitrification that in turn might have increased the nitrogen content in these soils.

Fyles and Mcgill (1987), who found that the nitrogen recovery in four- and seven-year-old revegetated sites was 76% and 79.5% respectively as against 40% in a one-year-old site. Aweto (1981) reported that the total nitrogen content of fallow soils increased by 33% over a period of ten years. Hossner and Hons (1992) recorded a 22% increase in total nitrogen content over a period of two years in the upper few centimeters of mines oils developing under selective species. Similar results of age-related increase in nitrogen content on restored mine soils in other sites (Soni and Singh, 2008).

The significant differences between the soils of reconstructed and undisturbed sites interms of their nitrogen content might be related to their differences in texture and organic carbon content, which are important in determining the rate of nitrification. Thus, the undisturbed soils with adequate organic carbon would have a higher rate of nitrification that in turn might have increased the nitrogen content in these soils. This assumption supports the findings of Secor et al. (1983), who indicated that high organic carbon and coarse texture of soils combined with moderate summer temperature and high oxygen tension would tend to favor the activities of nitrifying micro organisms.

In view of the importance of soil carbon to global carbon budget (Revelle and Munk, 1977 and Schlesiger, 1985), the accurate estimates of soil carbon pool shave assumed greater significance. Forest soils contain a conspicuous portion of world's carbon reserves (Zinke et al., 1984), but these pools are difficult to estimate quantitatively because of the high variation in soil properties normally compounded by many disturbances. A soil is considered organic if the content of organic carbon exceeds 18%. According to Troeh and Thompson (1993), 1% of organic carbon has about as much influence on soil.

The mobility of weathered uranium in the soil profile is dependent upon the affinity of the soil for uranium and properties of the soil, such as its acidity or alkalinity and water content. Thus, where soil has a high affinity for binding uranium and its release into soil (typically soils high in organic matter), water should be minimal. Correspondingly, mobility is likely to be greater in soils that bind uranium less strongly, which includes soils of semi arid climate where neutral to alkaline soil pH is combined with low organic carbon content (Ebbs et al., 1998). Plant uranium concentration was mostly affected by concentration of uranium in the soil.

The comparison of pH value of adjacent undisturbed soils with a revegetated site showed that the value was the same in forest land and village land. The low pH in native soils could be an indication of the presence of higher amounts of organic carbon and the fine texture of the sesoils. The soils on undisturbed areas that are highly weathered and leached are prone to loss of exchangeable bases to lower surfaces. These results support the findings of Indorante et al. (1981), who found the pH of soil to be higher on restore dare as than on the natural are as they studied. Higher pH values in recent plantations compared to adjacent natural forests has also been reported by Adejuwon and Ekanade (1988).

After nitrogen, the deficiency of phosphorus in most of the mine soils is universal (Bauer et al., 1978) and its availability depends on soil pH; with both high and low values leading to an immobilization of this element (Tabatabai, 1985). Although there is little indication that phosphorus is a limiting plant nutrient for native vegetation (Beadle, 1985), the anticipation of its low concentrations on degraded mine soils is of crucial importance to restore their fertility status (Munshower, 1993).

The age-related increase in the availability of phosphorus on restore dare as as well as its higher concen tration on undisturbed sites can beat tributed to the accumulation of organic matter; the phosphorus of which is converted to orthophosphate upon mineralization (Karimian and Razmi, 1990).

Fuller (1974) has stated that much of the phosphorus of native soils occurs in organic forms, and Tiedemann and Klemmedson (1973) suggested that higher levels of organic matter under some shrubs are indicative of an increased availability of phosphorus in these soils. Karimian and Razmi (1990) observed that the soils under range plants with the highest amount of organic matter also contained a higher concentration of available phosphorus. The highly significant correlation between the set wo soil factors further lends support to this assumption. Weathering and leaching which are at an advanced stage on an undisturbed site also appear to have initiated on the experimental area. Both of these processes have been shown to improve the phosphorus availability because organic forms and adsorbed ions are produced after removing the calcium gradually (Troeh and Thompson, 1993). There is also an indication that a limited amount of phosphorus converted to organic forms is adsorbed on the surfaces of other compounds such as octo calcium phosphate, which is more easily available for plant use (Olsen et al., 1983).

While a decrease in the concentration of exchange able potassium on revegetated areas may be ascribed to high leaching (Martin and Ryan, 2004), the low levels of exchangeable potassium in the soils of forest land and village land may have resulted from the rapid mineralization and humification of organic carbon. Soni et al. (1992) also found that severe weathering, high leaching and clay eluviation tend to reduce the potassium content of surface soil horizons.

3.2 RADIONUCLIDE STUDIES IN PLANTS, SOIL AND TAILINGS

The mobility of weathered uranium in the soil profile is dependent upon the affinity of the soil for uranium and properties of the soil, such as its acidity oral kalinity and water content. Thus, where soil has a high affinity for binding uranium and its release into soil (typically soils high in organic matter), water should be minimal. Correspondingly mobility is likely to be greater in soils that bind uranium less strongly, which includes soils of semi arid climate where neutral to alka line soil pH is combined with low organic carbon content (Ebbs et al., 1998). Plant uranium concentration was mostly affected by concentration of uranium in the soil.

The radionuclides, which are mobilized within the plant body, eventually move to human systems through various food chains. According to soil science literature, clay and organic matter are active components for providing nourishment to plants. They have strong cationic adsorptive capacity owing to the presence of negative surface charges. The variables that can affect the migration of radionuclides and trace elements from soil/substrate include pH, moisture content, cation-exchange capacity and competing ions (NCRP, 1992). Apart from this, the physical condition of the environment, such as temperature and soil textures, are also important in deciding the extent of bioaccumulation in plants. The radionuclides and trace elements follow a complex pathway before being associated with a metabolic system.

Different workers have carried out works on uranium mill tailings impoundment. Physical characteristics of the tailings' (particle size) leachability and associated radio activities have been reviewed by Markose (1990). Efforts were also made earlier (Webb and Voorhees, 1984) to evaluate the dispersal of tailing piles through various natural processes and engineering failures. Eisenbud (1987) has discussed the methods to reduce the radon emission through the application of soil cover over the tailings. Gans et al. (1981) have reviewed the migration characteristics of radionuclides from the tailing pile. Tsivoglue and O'Connel (1963) reviewed the nature and distribution of contaminants from the uranium ore processing industry. Movement of uranium in soil and vegetables has been reviewed by Morishima et al. (1976). Uptake of uranium by plants growing over uranium vanadium deposits of the Colorado Plateau has been reviewed by Cannon (1952).

Uranium mill tailings happen to be the chief carrier of ^{226}Ra because they contain the whole spectrum of radionuclides belonging to the ^{238}U decay series. As such, the disposal of processed waste from the uranium industry constitutes a vital link in the migration and buildup of "radium" in the abiotic and biotic components of the adjoining ecosystem. Evaluation of migration characteristics of radium is of concern owing to its high degree of radiotoxicity, long radiological half-life and metabolic link with calcium (a major nutrient) in humans. Moreover, several workers have reported dispersal of radium into an adjoining environment near a uranium industry. Uptake and translocation of radium to different organs of plants growing over exposed tailings have been extensively studied here and is the subject matter of present discussion. Uptake of radium by other plants growing in the adjoining aquatic ecosystem and vegetables growing on normal soil are also discussed. Site- and species-specific behavior of radium, the interrelationship of radium with other stable elements and an evaluation of concentration ratio from substrate to plant/plant organs is also a part of this investigation.

While the growing of vegetation and plants over remediated tailings has many advantages, there are a few disadvantages as well. Roots of vegetation create cracks and fissures, causing an increase in the porosity of the top layer. This may cause an increase in radon emanation and the percolation of rainwater to the tailings. Some plants pick up radioactivity and trace metals from the tailings or the cover material, which has a small potential of reaching the cattle and human food chain.

Vegetation cover of grass and shrubs not grazed by cattle, such as *Saccharum spontaneum* (Kans), *Typha latifolia* (cattail), *Ipomoea carnea* (Amari) and *Lantana* (putus) were grown over the used-up portion of the first two tailing ponds in Jaduguda. This has been found effective in consolidating the tailings and suppressing generation and dispersal of dust, besides merging it with the local landscape.

Uptake of radionuclides by these plants and shrubs, including the *Cynodon dactylon* (doob grass) has been studied (Basu et al., 2000). *Cynodon dactylon* shows the highest affinity for Mn and Zn, which are present in the tailings. This is the only component of the vegetation that may be grazed by cattle. All other species mentioned earlier showed a soil to plant transfer factor in the range of 10^{-3} to 10^{-2}. Typha latifolia showed relatively higher uptake. Laboratory experiments carried out on vetiver grass (*Vetiveria zyzanioides*, khas grass) have shown high uptake for uranium from uranium tailings (Venu Babu, 2007). After maximum uptake, the vetiver is reported by the researcher to dry up within 35 to 40 days. This appears to have an attractive phytoremediation property. This variety is also known for high soil binding properties.

The vegetation or plants that show significant uptake of radionuclides and trace elements may be suitable for phytoremediation over short periods of time, but they may not be suitable for a vegetative cover directly over the tailings due to the probability of radionuclides and trace metals reaching the food chain. However, due to their shallow root penetration and survival under harsh conditions, they are attractive candidates for plantation over remediated and soil-covered tailings. This is true for the vetiver (khas grass) as well (Venu Babu, 2007).

Some studies carried out in other countries have indicated that Indian mustard (*Brassica juncea*) and annual sunflower (*Helianthus annus*) could be suitable for phytoremediation, but studies carried out in Germany demonstrated that transfer from soil to plant, (especially the roots) was of the order of 10^{-3}(Huhle et al., 2008). Thus, their suitability for effective phytoremediation could not be confirmed in field studies. In view of low or negligible uptake of radionuclides and shallow root penetration, these may be used on the soil cover over the remediated tailings (Khan, 2008). Studies at the Indian Institute of Mine, Dhanbad, indicated that lemon grass (*Cymbopogon flexuosus*) could be grown on fly ash (Ghosh et al., 1998). As lemon grass also has shallow roots, its suitability for growing over remediated tailings may be studied. To reduce the dispersal of dust in to the environment, it is suggested that planting Jatropha, a source of biofuel, on the periphery of tailings facilities may be considered as a part of the greenbelt (Khan, 2008). According to UMH-IV, 2005 operations in eastern Germany have carried out large-scale remediation of waste rocks and uranium mill tailings over a very large area. (After successful remediation and vegetative cover, many sites have been released for floriculture, sports and recreational activities.)

A comparative uptake study of uranium and radium was carried out by Rumble and Bjugstad (1986) at the uranium mill tailings disposal site and control areas of South Dakota. A higher concentration of uranium was observed in plants growing on uranium mill tailings than in those from the control sites. Rickard and Keough (1968) suggested that plant uptake of uranium is independent of soil concentration. It was reported by the same author that the concentration of uranium in soil is greater than 200 ppm. These are toxic to some plants, and uptake of uranium from the soil by plants is in the range $10^{-1}10^{-2}\mu g.g^{-1}$ calculated on a fresh weight basis. The higher concentration ratios for uranium on the mill tailings would suggest that there is some dependence of uptake of these radionuclides on initial soil concentration. The long-term implication of radionuclide uptake involves the accumulation of uranium on the soil surface from many years of litter fall and decomposition, as well as off-site movement by wind and water.

Plant tissue concentrations of radionuclides have rarely shown a linear relationship to radionuclide concentration in substrate (Sheppard and Sheppard, 1985; Sheppard and Evenden, 1988; Sheppard et al., 1989). Plant and soil concentrations were not significantly correlated; the correlation coefficient between AE horizon soil and leaf concentration was −0.13, whereas it was −0.41for stem. Plots of CRs identify a clear differentiation by tissue type and by nuclide. Although CRs based on AE horizons are somewhat higher, CR trends were virtually identical for both AE and Bs horizons. CRU_{lv} values were not significantly different from CRUst. When CRs were calculated based on AE horizons, CRTh in both stem and leaves declined with increase in soil concentration. Additionally $CRTh_{AE}lv$ and $CRTh_{AE}st$ were highly correlated (R = 0.88, p = 0.001), but $CRU_{AE}lv$ values were poorly related to $CRU_{AE}st$. Studies by Sheppard and Evenden (1988) found that U and Th CR values decrease as soil concentration increases. Control pot study of blueberry plants grown in organic soil artificially contaminated with uranium were showing CR values nearly an order of magnitude higher as reported by Morten et al. (2002), although CR values decline significantly with time. This suggests plant uptake may eventually equilibrate. Neither uranium plant concentration nor uranium CRs were linearly related to uranium soil concentrations suggesting that soil uranium may be partitioned more explicitly into available and unavailable forms. Weathering, processes responsible for particle size reduction and increased surface area appear to enhance the release and availability of uranium. Studies carried out at the Anaconda mill site for forage uptake of uranium series radionuclides has given an uptake coefficient for ^{238}U (ratio of vegetation activity to soil activity) that varies from 0.69 in Indian rice grass roots to 0.01 in four-wing saltbush (*Atriplex canescans*) shoots (Rayno et al., 1980).

Uranium concentrations in some crops (high-yielding *Brassica* family, root crops such as beets, Swiss chard and sugar beet and high biomass crops like corn, sunflower, *Cynodon dactylon*, etc.) were evaluated by application of uranyl nitrate solution and EDTA complexing agent in the ratio (1:1). Plants were harvested after six to eight weeks, and the uranium in the dry matter content of various plant organs were determined. Generally, the addition of EDTA increased uranium concentration in the roots and shoots of the *Brassica* species 10% to 35%. For some plant species, the addition of EDTA has little effect on uranium accumulation. Root crops and high biomass crops differed in uranium uptake and accumulation. Some plants absorbed and translocated uranium to a greater extent than did others. Among the plant species tested, sunflower accumulated the most uranium, mainly in the root (800 µg/g dry). Sugar beet had the highest uranium concentration in the shoots and Bermuda grass and corn had the lowest. Except for Swiss chard and corn, the addition of EDTA significantly increased uranium concentration in shoots of all plants (Hossner et al., 1997).

Rufykiri et al. (2001) discussed that arbuscular mycorrhizal fungi can transport uranium under root organ culture conditions. An experiment related to the transport of uranium by arbuscular mycorrhizal fungi showed that uranium was mobilized by Arbuscular mycorrhizal fungi (AMF). Mycorrhizal carrot roots exposed to a low level of uranium mobilized more uranium per biomass unit than did nonmycorrhizal carrot roots. This AMF effect was not observed at relatively high uranium levels. Uranyl cation (UO_2^{2+}) is stable under acidic conditions, phosphate and hydroxyl complexes under acidic or near neutral conditions, while carbonate complexes predominate under alkaline conditions. Noteworthy was the high uranium transport to roots observed at pH 4, indicating that soluble uranyl cation or uranyl-sulphate species were more mobilized by fungal tissues, while phosphate species dominating at pH 5.5 or hydroxyl and carbonate species dominating at pH 8 were rather immobilized by hyphal structures. The high bioavailability of uranyl cation was also reported in another study involving plants (Ebbs et al., 1998). Such results suggest that pH is an important factor whose influence cannot be ignored when studying the mechanism of uranium bioavailability.

Jovanovic et al., (2004) has studied uptake and distribution of uranyl nitrate in soybean (*Glycine max*), sunflower (*Helianthus annus*) and maize (*Zea mays*) plants. The effect of uranyl nitrate application on plant physiology was also studied by the author for a definite follow-up period (12 hr–100 hr) after application. No serious damage for the plants was observed for the entire determination period after application of 150 and 300 ppm uranyl nitrate solution. However, some slight leaf damage was visible on sunflower plants after 100 hours. Uranium content in roots rapidly increased after 12 hours, nearly doubling after 100 hours in sunflower and soybean. On the other hand, the content of uranium in maize roots after initially increasing 20% continued to constantly decrease until the end of the experiment. The average uranium content in the entire plant was highest for sunflower, lower for maize and lowest for soybean (5056, 2264 and 1835 respectively in relative units). Uptake rate calculated from the first part of the uptake curve was the highest in sunflower, lower in soybean and the lowest in maize. To conclude, sunflower was found to be highly suitable for uranium remediation of nutrient solution, as its uranium uptake is fast and the amount taken up is high.

Gunther et al. (2003) studied the uranium speciation after uptake of uranium by lupine plants. Differences were detected between the uranium speciation in the initial solution (hydroponic solution and pore water of soil) and inside the lupine plants. The oxidation state of uranium did not change and remained hexavalent after it was taken up by the lupine plants. The chemical speciation of uranium was identical in the roots, shoot axis and leaves and was independent of the uranium speciation in the uptake solution. The results indicate that the uranium is predominantly bound as uranyl (VI) phosphate to the phosphonyl groups.

Knowledge of the levels of the naturally occurring radionuclides is important because they contribute a substantial portion of the radiation dose to humans, and diet is usually the major source of these radionuclides (NCRP, 1975; UNSCEAR, 2000).

210Po (T1/2 = 138 days), an alpha emitter of the 238U series, is of radiological interest for a number of reasons, but mainly because of its large contribution to the internal radiation dose (Skwarzek, 2000). Evidence from the literature indicates that 210Po is accumulated strongly by organisms and transferred via food along the trophic chain (Heyraud and Cherry, 1979; Skwarzek and Falkowski, 1988; Khan et al., 2002). Because of the possibility that large amounts of 210Po may be thus transferred to humans via diet, there is a need for the investigation of this nuclide in foodstuffs. The concentration of 210Po has been evaluated in some of the locally growing fruits and vegetables around the proposed uranium mining sites of Bagjata and Banduhurang (Jha et al., 2007).

With regard to environmental protection, long-lived radionuclides of ^{238}U decay series are a prime cause of concern *viz.* ^{230}Th, ^{226}Ra, ^{210}Pb and ^{210}Po. They have half-lives of 80,000 years, 1622 years, 22 years and 138 days respectively. Owing to the low abundance of ^{235}U in the ore (0.7%), its daughter products are of less significance for environmental protection.

The two most significant radionuclides that may appear in the environmental matrices are U (nat) and ^{226}Ra in view of the waste fixation technology. The former is chemically more toxic and accumulates in the kidney, whereas the latter is a bone seeker with behavior similar to calcium (a known essential element for bone development). Consequently, from the point of view of abundance, established chemical/radiological toxicity and long half-lives, these two elements are the most significant radionuclides in the uranium processing industry. In the text that follows, the terms "uranium" and "radium" refer to natural uranium and ^{226}Ra unless mentioned otherwise.

Uranium contamination can arise from mining activities, fuel fabrication, fuel reprocessing plants and effluents from both mining and milling plants. While remediation of soil and effluents is a mandate from the standpoint of public health, recovery is desirable, as India has only a limited natural resource of uranium. Globally, several studies employing both microbially driven bioremediation techniques as well as phytoremediation efforts were carried out, and a wealth of information is already available. While great advances have already been made in areas like bioleaching (Mathur, 2005), recovery schemes of uranium from contaminated soil and effluents are still far from perfect for routine use.

The movement and subsequent off-site dispersal of radionuclides are essential factors for risk assessments and determining remedial actions at hazardous, radioactive waste sites. One potential transport path from the soil column is through the uptake and deposition of plant material to surface soil. The behavior and mobility of radionuclides in soil is a major consideration for plant uptake and is influenced by many variables (Romney et al., 1974; Poston et al., 1995a). The uptake characteristics of radionuclides have been observed to vary depending on soil properties such as texture, organic matter content, bacterial action, pH, redox potential and physiochemical speciation (Poston et al., 1995b; White and Dunaway, 1975; Romney et al., 1974; Han and Lee, 1997; Salbu and Oughton, 1995). Nuclear energy has already started playing a significant role in civilian economics, in view of the rapid depletion of reserve fossil fuels. Therefore, there has been an emphasis to increase and augment uranium production round the world. The increased uranium production will result in the generation of processed waste, which in turn will pose challenges for its safe disposal on long-term basis. This is because such wastes are not only radioactive, but also chemically toxic.

Poston et al., (1995a) evaluated the concentrations of radionuclides in soil and vegetation samples collected on-site and off-site as documented by the Sitewide surveillance program. The radiochemical analyses used to monitor radionuclides in soil and vegetation were extremely sensitive, and past monitoring has documented the accumulation of low levels of radionuclides at several areas. Atmospheric fall-out from past nuclear weapons testing programs may also contribute to levels of radionuclides analyzed in environmental media collected for site surveillance. The report specifically addresses 0Co, 9 0Sr, 1 3 7Cs, U isotopes, 2 3 8Pu, [239,240,241]Am, in surface soil and vegetation samples. Gamma spectroscopy provides information on a number of other radionuclides; however, these are consistently below the level of detection.

While microbial systems with promising mechanisms for dealing with heavy metals are the subject of contemporary studies, many plants, including crop plants, grasses and a wide variety of species with hitherto unknown usefulness to human kind, have been demonstrated to be cheap alternatives in waste management systems. The latter approach, called phytoremediation, involves various processes (Yang et al., 2005) such as phytoextraction, rhizofiltration, phytostabilization, phytovolatization and phytotransformation.

Several plants, called hyperaccumulators, have been identified that are capable of accumulating metals up to 1% in shoots, when challenged by a deficiency of nutrients (Minguzzi and Vergnano, 1948; Yang et al., 2002). Members of Brassicaceae (*Alyssum* sp., *Thlaspi*), Euphorbiaceae (*Phyllanthus* sp., *Leucocroton* sp.), Asteraceae (*Senecio* sp., *Pentacalia* sp.) and Lamiaceae (*Aeollanthus* sp.) are known hyperaccumulators of Ni, Zn, Cu and Co (Suresh and Ravishankar, 2004). The mechanisms involve uptake of metals by plants possibly through apoplast and/or symplast assisted by ligands such as citric, oxalic or malic acids secreted by roots and by metal transporters expressed in root cell membranes. The metal ions are either complexed with metallothioneins (sulfur-containing metal-binding proteins) in cytoplasm or sequestered by phytochelatins (short, nontranslationally made peptides) for storage in vacuoles within the root cortex. They are also mobilized into stele and are loaded into xylem, where certain ligands like histidine are complexed and the metals are transported to shoots. The distribution of metals within the leaves once again involves transporters, phytochelatins and so on, and the metals are utilized in metabolism or are either sequestered for storage in vacuoles in epidermal cells or delivered into special organs, such as trichomes for secretion or compartmentalization, to safeguard cells from damage (Suresh and Ravishankar, 2004; Yang et al., 2005).

Several genes from bacteria, mouse, yeast and plants have been already mobilized into model plant systems, using a transgenic approach, and were shown to be effective against pollutants like Hg, Cd, Pb, Ni, Al, Zn and As (Eapen and D'Souza, 2005).

Uranium is found in all rocks and soils. As high as 120 ppm is found in phosphate rocks of Florida and southeastern Idaho. The neighboring areas can be used as commercial sources of uranium (Clegg and Foley, 1958; NCRP, 1975). The high uranium content of phosphate rocks is reflected in corresponding elevated levels in phosphate fertilizers.

4 Species Selection for Revegetation and Consolidation of Uranium Tailings

Ecological risk assessment and management of land consolidation are two objective requirements of the sustainable development of land resources under the background of ecological civilization. Monitoring and controlling ecological risks has become an issue gaining more and more attention. On the basis of systematic elaboration of the connotation and characteristics of ecological risks of the land consolidation, analysis has to be done of ecological risk factors and evaluation index system relative to the ecological effect of land consolidation, and thereof, a new idea was brought forth about identification, assessment and control of ecological risks based on full-process analysis of land consolidation projects, including ecosystem service functions as an important content of ecological risk assessment. Thus, the limitation of research in the past on ecological risk has been overcome with more emphases on changes in the structure and functions of an ecosystem. Scientific issues and development trends that need more attention in tackling ecological risks of land consolidation are pinpointed (Liu et al., 2017).

A special study on vegetative covers was recently carried out as part of the Uranium Mill Tailings Remedial Action Project (UMTRAP) in the United States (US Department of Energy, 1988). This study investigated the use of vegetation to stabilize uranium tailings, and specifically includes the use of vegetation to intercept infiltration. The study showed properly developed plant communities on complex soil covers can be effective in stabilizing covers and controlling infiltration on top slopes of waste piles. Further, the appropriate vegetative cover will adapt to climatic change, will repair itself after severe disturbances such as fires and drought and will persist indefinitely with little or no maintenance. The plants were found to protect top slopes against sheet wash erosion; however, resistance to gully erosion depends more on the overall pile configuration than on the vegetation and soil.

Rehabilitation is the process by which the adverse impacts of mining on the environment are repaired. It is an essential part of developing mineral resources in accordance with the principles of sustainable development. Minerals are a major source of raw materials in industries, and therefore mining is inevitable. In the process of open cast mining, several changes occur in the physical, chemical and microbiological properties of soils (Soni et al., 1994).

DOI: 10.1201/9781003326151-4

4.1 ECOLOGICAL STRATEGY AND CRITERIA FOR SELECTION OF SPECIES

Ecologically viable species should be selected for remediation of uranium mill tailings, and selection has to be based on the criteria of lowest demand for inputs and negligible attention needed. The selected species should have a deep and large root system and preferably be hardy, fast growing and suckering (Sastry and Kavathekar, 1990). Native vegetation could be more useful and efficient in restoration and bioengineering programs, as they have fewer competitors than other species and the disturbances permit the germination and development of nonseeded species (Munshower, 1993). The role of native vegetation is quite significant in preventing landslides, as well as for stabilizing the landslide. However, for remediation of mill tailings, plant species with a shallow root system but with good binding capacity have to be preferred. Plant cover not only protects the surface from weathering due to the direct impact of rain and winds, but its intricate root system works as a cohesive in binding the loose soil and preventing it from erosion (Rawat et al., 2012).

Pamela Graff and Sue, M (2014) studied plant attributes under three different restoration scenarios (high-level conservation, functional native vegetation and perennial native pasture) and generated prioritized lists for restoration of grassy woodland species. They used the following methodology for selecting plant species for restoration, which harnesses expert opinion on the relative importance of ecological criteria (including some traits). The process involved mainly the following steps:

1. identification of a *plant assemblage* for which restoration is considered important;
2. creation of three *restoration scenarios*, characterized by land use histories and management; and
3. selection of *ecological attributes* that might have relevance in determining the establishment and persistence of a species at the range of receiving sites.

Considering the aforementioned scenarios, the following seven ecological criteria (and their attributes, in italics) were selected as being relevant to the establishment and persistence of species. The rationale is given for each choice. Strictly speaking, three of the criteria are biological traits (*dispersal, life cycle, seed availability*), two relate to distribution patterns (*geographical range, conservation status*) and two are ecological responses (*phosphorus tolerance* and *grazing tolerance*).

It is a common practice to treat contaminated sites with traditional methods like physical, chemical and thermal processes resembling excavation and transportation (Singh et al., 2014). Several chemical treatments are currently available to treat uranium-contaminated sites, but these methods require expensive extraction of the soil and exposure of workers to uranium (Gavrilescu et al., 2009). Traditional physical methods involving removal of 1 m³ soil from a one-acre contaminated site is estimated to be US $0.6–$2.5 million (McIntyre, 2003).

Phytoremediation in its general sense means cleaning contaminated sites with appropriate plants, which offers not only greater potential to remediate contaminated sites over conventional and costly methods, but also a means to use a "green", sustainable process (Pulford and Watson, 2003; Niu et al., 2007).This type of process generally has high public acceptance and can often be carried out on-site to remedy contaminated soils, sediments and/or groundwater (Vidali, 2001; Dietz and Schnoor, 2001). Except for a few limiting factors, this technology has the ability to rejuvenate contaminated environments effectively. Due to its approach, it is an environmentally friendly (eco-friendly), safe and cheap way to clean up contaminants or pollutants and has aesthetic advantages and long-term applicability (Schnoor et al., 1995). Phytoremediation can be used at sites containing organic, nutrient or metal pollutants that can be accessed by the roots of plants and sequestered, degraded, immobilized or metabolized in place. In the last few years, greater understanding has been achieved of this process.

Good planning and environmental management minimizes the impacts of mining on the environment. Bradshaw et al. (1975) found that nitrogen accumulation is a major factor limiting the rate of vegetation development. The universal deficiency of phosphorus in most of the mine spoils have been reported by various workers, who stated that the specific value of pH acts as an ecological indicator of plant species (Power et al., 1978). Soni (1995), while studying biorestoration of surface mined lands, had suggested that successful biorestoration of derelict surface mined lands depends on assessment of site-specific problems, selection of appropriate plant species to ensure revival of the ecosystem and application of a package of practices suitable for the site.

In light of the few available reviews in the field of Indian restoration ecology (Ghose, 2004a; Ghose and Kumar, 2004; Prasad, 2007; Soni et al., 1989), a number of knowledge gaps have been identified regarding the broadscale depiction of post-mined land rehabilitation in India. Among others, a few lingering research questions deserving of further consideration include the following: Which plant species have been studied for the purpose of land revegetation? Which of these are considered most suitable for determining revegetation success, and what characteristics make them more appropriate than others for this purpose? How are environmental and (or) physiological factors used to determine revegetation success? What is the significance of these factors toward ecosystem restoration?

Native vegetation is grown directly on most mine wastes to consolidate and stabilize the rocks and dumps. Even on some abandoned uranium tailings in remote locations in Canada and some other countries, where human or animal access is less probable, the local shrubs and plants are grown directly over the tailings as part of the remediation measure. But growing vegetation directly over uranium mill tailings near populated areas may be undesirable due to the probability, albeit low, of transfer of radioactivity to the animal and human food chain and associated adverse public opinion (Khan, 2008).

While the growing of vegetation and plants over remediated tailings has many advantages, there are a few disadvantages as well. Roots of vegetation create cracks and fissures, causing an increase in the porosity of the top layer. This may cause an increase in radon emanation and the percolation of rainwater to the tailings. Some plants pick up radioactivity and trace metals from the tailings or the cover material, which has a small potential of reaching the cattle and human food chain.

4.2 ROOT PENETRATION AND SOIL BINDING CAPACITY

Most plants take up their nutrients and contaminants mainly through the roots from the soil solution, although absorption through leaves also occurs. The extent to which uranium is bound to soil components, and the strength of that binding, affects the amount of soluble soil uranium available for uptake into the plants. Therefore, the factors influencing uranium mobility in soil are also likely to exert a strong influence on the extent of plant contamination. The uptake of uranium by plants, although low compared with mobile radioactive elements such as cesium and strontium, is higher than that of plutonium and americium. The soluble forms of uranium seem to be readily absorbed by the plants. *H*owever, in many soils, natural uranium has a low solubility and can be unevenly distributed. In general, uranium concentration in plants decline in the order: root >shoot > fruit and seeds. A stimulatory effect on growth has been observed in some grasses exposed to elevated concentrations of uranium in soil at broad concentrations. The concentration ratio has been determined for different matrices: mine waste, tailings, processed waste and so on (Royal Society, USA, 2000). It has been pointed out that the concentration ratio varies by four orders of magnitude for crops on different soils and different soil uranium. This wide variation severely inhibits the applicability of generic models that incorporate uptake by plants and highlights the need-based approach with well-defined source terms and composition (Rodriguez-Freire et al., 2021).

The below ground competition and greater responsiveness of roots on a restored site can be seen to influence the stability of soils in the early stages of development (Nambier, 1981). The values of total root length recorded on both sites are well within the ranges recorded by Nambier (1983). In view of this large network of roots, it appears that shrubs should occupy much of the soil volume, and thus their importance in soil physics is obvious. It is this network of a large quantity of roots that has dramatically changed the edaphic factors on dry and infertile soils of a mine site. The observations of improvement in soil characteristics with the addition of organic matter from root decay have been reported by several workers (Thakur, 1957; Bhimaya and Seri, 1980; Singh and Gupta, 1983).

Generally, the greater root length in natural habitat soils may not be of considerable importance compared to its significance for plants grown on degraded habitats, where soil moisture can be considered the main factor to inhibit plant survival and growth. Thus, the aforesaid species with more lateral roots as well as greater root surface area have an advantage to absorb more water. With this character, these species have a higher capability of growing over stress sites. Since absorption of nutrients also depends on root length and root surface area, it can be assumed that these species with a proportion of branched roots would absorb more minerals per unit of weight than would unbranched roots (Nye, 1966).

5 Plant Growth Studies

Plant growth is known to improve the heterotrophic microbial community, which also promotes metal stabilization through complexation or solubilization and improves plant growth. Therefore, instead of growing a single species for phytostabilization, it is always beneficial to come out with a community of plants that can be promoted to grow over mill tailings. As such a mixture of different plant species promotes growth of different microbial species that can complement the process of complexation or sequestration in the rhizosphere. The extensive growth of plants was certainly due to the presence of an extensive network of fibrous roots (Masoodi 1998). The major portion of the roots was found in the upper 25 cm of soil. The data are in agreement with the findings from other studies (Schafer et al., 1980; Roberts et al., 1988). This led to an increase in organic carbon content in the upper surface layer of restored soil capping on tailings. The increase in organic carbon in the restored area may be ascribed to the addition of plant litter and other residues to the soil surface. Further, the growth pattern of different species also uses the soil moisture to optimal levels while providing the best vegetation cover to the soil caps all round the year. However, it is imperative to follow certain cultural practices in order to establish a community of plants for the purpose of phytostabilization.

Many plants alter soil and soil biota, which in turn may alter the formation of soil and plant growth. According to Wardle et al. (2004), there are two main pathways on which the interaction between plant and soil operates: (i) associated with roots and (ii) associated with litter. Plant-soil feedback part associated with roots and root exudates comprises interactions with distinct root symbionts and another rhizosphere biota (Bonkowski et al., 2000). This biota may have both an advantageous and an adverse impact on plants (Hinsinger et al., 2003; Jones and Smith, 2004). The beneficial organisms mostly consist of fungal symbionts and symbiotic nitrogen-fixing bacteria (Smith and Read, 2008). Plant pathogens, parasites and herbivores may accumulate in the rhizosphere and have a negative impact on the plants (Kardol et al., 2006; Cortois et al., 2017). On the other hand, root-associated microbes such as mycorrhizae considerably influence the growth of plants and subsequently the plant community (Nara, 2006) and herbivores (Fagan and Bishop, 2000). The class of interactions that results in plant-soil feedback is connected to plant litter and its consequences on release of nutrients and topsoil formation (Ponge, 2013; Frouz, 2018). Litter characterizes mainly C and input of nutrients into the soil. It is usually recognized that fast-growing plants with thin leaves are responsible for the production of litter. They are highly rich in nutrients, decompose rapidly, and are responsible for the release of many nutrients compared to plants with conservative growth methods (Cornwell et al., 2008).

Globally, two main concerns are the increase in the chain of contaminated or damaged lands and high bioenergy demands. Hence, linking phytoremediation with energy crops is needed now and in the future for sustainable development. Many countries are developing the potential of phytoremediating contaminated soils while also analyzing the potential of planting various energy crops for the production of biodiesel (Meers et al., 2007; Olivares et al., 2013; Pandey et al., 2015). Highly promising energy crops are cultivated for various benefits on contaminated lands. Utilization of degraded, damaged, contaminated lands has become the partial solution for the problem of restricted agricultural lands as well as advantageous from an economical point of view (Gopalakrishnan et al., 2011). Reuse of poorly vegetated contaminated sites for bioenergy crop production might be an advantageous solution (Kechavarzi and Lord, 2009). Maximum work has been carried out on fast-cultivating woody plants for phytoremediation of contaminated lands (Punshon et al., 1996). Energy crops are usually suggested for wasteland degradation to avoid competition with food crops. All types of contaminated lands have low levels of fertility due to poor soil that cannot cultivate the adequacy of plants for optimum production (Pandey et al., 2012). To measure the bioproductivity of a natural ecosystem or a restored site, the component of plant growth that is important is the net primary production or total yield. Two types of measurements are needed for growth analysis: (1) the plant weight—usually the oven dry weight (kg) but can be the organic matter or energy content—and (2) the size of the assimilatory system—usually the leaf area (m^2) but can be the leaf protein or chlorophyll content.

Crop canopy spatial parameters are indicative of plant phenological growth stage and physiological condition, and their estimation is therefore of great interest for modeling and precision agriculture practices. Rapid increases in computing power have made stereovision models an attractive alternative to common single-image-based 2D methods, by allowing detailed estimation of the plant's growth parameters regardless of imaging conditions. Models that have been proposed thus far are still limited in their application because of sensitivity to outdoor illumination conditions and the inherent difficulty in modeling complex plant shapes using only radiometric information. Assuming that not all of the plant-related pixels are essential for growth estimation, this study proposes a 3D reconstruction model that focuses on selected salient features on the plant surface, which are sufficient for obtaining growth characteristics. In addition, by introducing a hue-invariant model, the proposed algorithm shows robustness to diverse outdoor illumination conditions. The algorithm was tested under greenhouse and field conditions on corn, cotton, sunflower, tomato and black nightshade plants, from young seedlings to fully developed plant growth stages, and accurately estimated height (error ~4.5%) and leaf cover area (error ~5%). Furthermore, a strong correlation (r^2 ~0.92) was found between the plant's estimated volume and measured biomass, yielding an accurate biomass estimator in the validation tests (error ~4.5%). This estimation ability remained stable while applying the model on plants with varying densities (overlapping leaves) and imaging setups where the standard 2D-based analyses failed, thus showing the 3D modeling contribution to robust growth estimation models.

Ecological restoration of mine tailings by revegetation provides a model system for investigating biotic interactions below and above ground. Microorganisms have been shown to rapidly respond to environmental change during the restoration process in an ecosystem of tailings. Micro organisms may be possible bioindicators for monitoring soil ecosystem functions in close association with changes in the physicochemical and biological conditions during the ecological restoration of mine tailings.

Vegetation such as grass, shrubs and plants are known to consolidate soil and prevent erosion. Hence, growing different types of vegetation is a normal practice to prevent soil erosion from embankments on dams and near elevated railway tracks and highways. Native vegetation is grown directly on most mine wastes to consolidate and stabilize the rocks and dumps. Even on some abandoned uranium tailings in remote locations in Canada and some other countries, where human or animal access is less probable, local shrubs and plants are grown directly over the tailings as part of the remediation measure. But growing vegetation directly over uranium mill tailings near populated areas may be undesirable due to the probability, although low, of transfer of radioactivity to the animal and human food chain and associated adverse public opinion (Engelbrecht, 2020).

The belowground competition and greater responsiveness of roots on a restored site can be seen to influence the stability of soils in the early stages of development (Nambier, 1981). The values of total root length recorded on both sites are well within the ranges recorded by Nambier (1983). In view of this large network of roots, it appears that shrubs should occupy much of the soil volume, and thus their importance in soil physics is obvious. It is this network of a large quantity of roots that has dramatically changed the edaphic factors on dry and infertile soils on a tailing site. The observations of improvement in soil characteristics with the addition of organic matter from root decay have been reported by several workers (Thakur, 1957; Bhimaya and Seri, 1980; Singh and Gupta, 1983).

Generally, the greater root length in natural habitat soils may not be of considerable importance compared to its significance for plants grown on degraded habitats where soil moisture can be considered the main factor to inhibit plant survival and growth. Thus, the aforesaid species with more lateral roots as well as greater root surface area have an advantage to absorb more water. With this character, these species have a higher capability of growing over stress sites. Since absorption of nutrients also depends on root length and root surface area, it can be assumed that these species with a proportion of branched roots would absorb more minerals per unit of weight than would unbranched roots (Nye, 1966).

High temperature and low moisture of surface mine spoil limit plant growth (Bradshaw et al., 1975) and also reduce decomposer activity. Wieder et al. (1983) reported that root development is restricted due to changes in physical and chemical properties during spoil handling. Lim et al. (1981) suggested that when sand content in the mine spoil was greater than 70%, it retained insufficient water for plant survival and growth. Concerning the revegetation method, Lim et al. (1981) reported on the strong acidic spoils in Taiwan. They found that species like *Ficus formosana*, *Rhus chinensis* and *Axonopus affinis* can be used as pioneer plants for the restoration of strongly acidic mine spoil.

Soil characteristics of tin mine tailings are generally low fertility and unfavorable physical conditions for supporting plant growth resulting from heavy disturbance during tin mining. Management of the tailings for further uses may be dependent upon the ownership of the tailing land. Attempts to utilize the tailings for agricultural use through experiments and trials have been made by the Tin Mine Spoil Research Station, among others. These included forest tree planting, growing of agricultural crops and improving the soil by several methods to elevate the fertility status and create a more favorable condition for plant growth. Successful methods consist of using chemical fertilizer either directly or by mixing with manure and other materials that are likely to improve the physical properties of tailings and ameliorate toxic conditions. There is still no large-scale application of the research findings due to difficulties in initial improvement of the topographic condition of the tailings and the economic situation of the owner. Future studies on mine tailings aim at lowering the cost of reclamation by using locally available materials to improve both physical and chemical conditions through the recycling of organic wastes and to improve planting methods to gain more plant production (Tanpibal and Sahunalu, 1989).

Ecological restoration, using tolerant plant species and nutrient additions, is a low-cost option to decrease environmental risks associated with mine tailings. An attempt was previously made to establish such a vegetation cover on an abandoned tailings facility in southern Ireland. Historically, the tailing site has been prone to dusting and is a potential source of contamination to the surrounding environment. The site was examined to determine the success of the previous restoration plan used to revegetate the site and to determine its suitability for further restoration. Three distinct floristic areas were identified (grassland, poor grassland and bare area) based on herbage compositions and elemental analysis. Surface and subsurface samples were taken to characterize tailings from within these areas of the tailing site. The pH of bare surface tailings (pH 2.7) was significantly more acidic ($p < 0.5$) than in other areas. Additionally, negligible net neutralizing potential resulted in the tailings being hostile to plant growth. Total metal concentrations in the tailings were high (c. 10,000 mg kg^{-1} for Pb and up to 20,000 mg kg^{-1} for Zn). DTPA-extractable Zn and Pb were 16% and 11% of the total amount, respectively. Metal content in grasses growing on some areas of the tailings were elevated and demonstrated the inability of the tailings to support sustainable plant growth. Due to the inherently hostile characteristics of these areas, future restoration work will employ capping with a barrier layer (Courtney, 2012).

According to Belcher et al. (1995), the competition intensity increased with biomass and became constant after reaching a certain level where competition intensity was considered to be the maximum within the system. At this point, where biomass was still too low and occurred on a restored site, shoot competition may not be as important as root competition. It is due to this competition that the plants on a restored site have allocated more carbon to roots so as to yield higher root to shoot ratios. In other study, Joshi (1995) recorded an aboveground biomass production of 3600 kg ha^{-1} and belowground production was 1930 kg ha^{-1} in some temperate grassland.

Singh et al. (1995) evaluated the effect of NPK fertilizer annually on growth performance of *Dendrocalamus strictus*, which was a suitable species for checking soil erosion and for revegetation of coal mine spoil. Native leguminous species show greater intervention in soil fertility parameters compared to native non leguminous species and also suggested that native leguminous species were more capable in bringing out differences in soil properties than exotic leguminous species in the short term.

Wilson et al., (1996) showed that species richness increases with an increase in the phosphorus status of soil. Fetcher et al., (1996) has reported that some species were able to grow faster than other species on soils and spoils low in nitrogen levels by the maintenance of leaf area. Vasistha et al., (1996) reported that plants grown in nursery soil have better higher growth as compared to those grown in mine spoils. Root/shoot ratio and diameter were observed to be greater in mine spoil. They also reported that mine spoil–grown plants have a higher collar diameter growth, root elongation and root diameter.

Rodrigues (1997) reported 64 tree species existing in the iron ore mine lands. This study helped in identifying potential tree species that have survived in the inhospitable conditions of the mines. These selected species can therefore be extensively used in the revegetation programs of the iron ore mine wastelands. Kumar et al. (1997) studied the community structure and biomass of vegetation in a ten-year-old ecologically restored rock phosphate mine. The study revealed that most of the species planted initially to restore the areas have been replaced by naturally invading species. The restored plant community has herbs, grasses, shrubs and trees. They came up with the conclusion that total aboveground biomass after ten years of ecological restoration is 43.94 tonnes/ha and trees contribute 93.58% to this.

Dadhwal (1999) observed the survival, growth and biomass production of 13 native species of shrub and grass species in the mine spoil for testing the relative performance of tree, shrub and grass species and reported that these all are based on biomass yield. Among trees *Leucaena leucocephala*, among shrubs *L. carnea* and among grasses *P. purpuream* and *Arundo donax* are the ideal choices for the rehabilitation of mine spoils. *Agave americana*, *Eulaliopsis binata* and *Chrysopogon fulvus* are also among the more suitable choices for the reclamation of limestone mine spoils based on soil binding characteristics and other uses.

Work done by Clewell (1999) on restoration of phosphate-mined land during which *Acer*, *Fraxinus*, *Ilex*, *Liquidambar*, *Magnolia*, *Persea*, *Quercus*, *Taxodium* and *Ulmus* shared dominance with short-lived volunteer willows (*Salix caroliniana*) that had already begun to senesce, fresh plantation of same species was done. The tree canopy exhibited 85% cover, and some trees had grown to 12.5 m tall. Basal area reclaimed 8.31 m²/ha for trees, 10 cm or more in diameter at breast height. Ten planted tree species produced seeds and yielded seedlings. The floristic composition over the decade consisted of 22 species of trees and 208 shrubs, vines epiphytes, ferns, graminoides and forbs. Thirty-eight nonarboreal species were directly transplanted, others arose from a seed bank in mulch that was amended on wetter sites and the rest volunteered via natural dissemination. The frequency of nonarboreal plants was collectively 98%. Seventy-three species at the restoration site were characteristic of the mature undisturbed reference ecosystem.

Panwar and Bhardwaj (2000) in their work attempted plantation of shrubs for rehabilitation of sandstone and limestone mined site. Three species, namely *Elacagnus umbellata*, *Coriarianepalensis* and *Indigofera pulchella*, were chosen for rehabilitation purposes. Among the three species planted, *Elacagnus umbellata* was observed to have maximum survival growth performance over *Coriaria nepalensis* and *Indigofera pulchella*, in both sandstone and limestone mined areas. In sandstone it registered 82.50% survival, whereas in limestone it was 91.37%. Addition of 2.5 kg forest soil per pit increased the survival and growth performance in both of the mines, except that of diameter and root/shoot ratio, which as observed was greater in mine spoil.

Healey and Gara (2002) examined the ecological effects of establishing a teak (*Tectona grandis*) plantation on an abandoned pasture in southwestern Costa Rica. The ten-year-old plantation's understory was evaluated using a nearby nonplanted area as a baseline for potential local recruitment of native tree species. The native trees in the teak plantation were significantly less abundant, less diverse and more restricted to the lower height classes than the trees in the abandoned pasture. Furthermore, trees with shrubby growth forms dominated the plantation's understory, whereas larger species were more common in the unplanted abandoned pasture. Several aspects of teak's biology may contribute to the species exclusion of native trees.

According to Vasistha and Negi (2005),not all species of plants can grow on these nutritionally and microbiologically impoverished areas. However, certain plant communities were adapted to colonize on mined sites. Understanding the process of natural colonization and plant succession in disturbed sites is a prerequisite for expedient restoration and conservation of biodiversity, since plant species that appear on abandoned sites provide a source of potential species for ecological restoration of mine-degraded ecosystems. They further concluded that colonization of the plant species in six-year-old abandoned limestone mined areas in the Mussoorie hills of the western Himalayas has been studied. A total of six grasses and herbs, 15shrubs and six tree species in the form of seedlings and saplings were found to be colonizing in these areas. *Eulalia* sp. and *Bidens biternata* among herbs and grasses, *Rumex hastatus* among shrubs and *Wendlandia exserta* among tree species were found to be most dominating species.

Singh (2005a) carried out a study to assess the degree of internal cycling of nitrogen and phosphorus in certain young native tropical wood species growing at two different sites of varying fertility status on mine spoil. The tree species, for example *Albizia lebbeck*, *Phyllanthus emblica*, *Pongamia pinnata* and *Dendrocalamus strictus*, were planted at two different sites. The researchers concluded from this attempt that nitrogen resorption was more efficient in plants growing in more fertile habitats than less fertile habitats, while no influence of the variation in habitat fertility on phosphorus resorption efficiency was shown. The leguminous species were more efficient for phosphorus resorption while non leguminous species were more efficient for nitrogen resorption.

Singh (2005b) evaluated the influence of NPK fertilization on biomass yield of tropical herbaceous legume *Stylosanthes hamata* sown for revegetation of coal mine spoil. It was concluded that NPK fertilization increased biomass production in herbaceous legumes, which in turn would improve the fertility status of spoil, allowing the colonization of native species, thus resulting in the quick establishment of a self-sustaining ecosystem.

Although plants devote a large proportion of their resources to roots, we have a poor understanding of the constraints under which root systems function. Roots are much less variable morphologically than leaves, and it is likely that root systems rather than individual roots are the focus of natural selection. In other words, architecture is more important than morphology. Existing classifications of root systems, based on the developmental model, have failed to provide much insight into their functioning and an alternative, topological model is outlined, in which the link is the basic unit of classification. Other components of the architecture of root systems, including link lengths, branching angles and diameters, and the ecological implications of variation are considered (Matthew et al., 2001).

Plant form and function have been discussed by (Robson et al., 2015), while Chapman and Lemaire (1993) have highlighted them orpho genetic aspects of plant growth and of root form and function (Dawson et al., 2000).

Yang et al. (2003) showed a field experiment at Lechang Pb/Zn mine tailings of Guangdong Province, southern China, to analyze the performance of growth and accumulation of metal in *Vetiveria zizanioides* and two leguminous species of plants—*Sesbania sesban* and *Sesbania rostrata* grown on fertilizer amended with tailings (also domestic refuse). It was stated that when using fertilizer alone and in combination with artificial fertilizer and domestic refuse, the growth and survival rates of *V. zizanioides* and two *Sesbania* species were enhanced significantly, specifically in combination. Roots of these species showed a similar concentration of heavy metals, while the shoots of *Sesbania* species accumulated three to four times greater concentrations of Cd, Cu, Pb and Zn compared to shoots of *V. zizanioides*. The metal uptake ability of Caryophyllaceae species (genera *Dianthus*, *Minuartia*, *Scleranthus* and *Silene*) from metalliferous soils in northern Greece with varying concentrations of Ca, Cd, Cr, Cu, Fe, Mg, Ni, Pb and so on were studied by Konstantinou and Babalonas (1996). It was concluded that *Scleranthus perennis subsp. Perennis* exhibits maximum concentration of Cu with 205 mg kg^{-1}, while *Minuartia cf.bulgarica* showed the highest accumulation of Pb at 1175 mg kg^{-1}. Concentration of Ca in plants was much higher compared to that of soil, while the contrary was true for Mg. In almost all cases, the ratio of Ca to Mg was less than one in the soil and greatly improved in the plants. For successful field phytoremediation, the choice of plant material is an important aspect. A field experiment conducted by Zhuang et al. (2007) to assess the phytoextraction of six biomass plants, namely *Dianthus chinensis*, *Rumex crispus*, *Rumex K-1*,*Vertiveria zizanioides*, and two populations of *Rumex acetosa*, compared to *Viola baoshanensis* and *Sedum alfredii*, which are called hyperaccumulators of metal. Paddy fields contaminated with Cd, Pb and Zn were used to carry out the experimental study, and the total extraction was observed to be 0.17 kg ha^{-1} and 32.7 kg ha^{-1}for *S. alfredii and V. baoshanensis* respectively. Maiti et al. (2005) conducted an experiment to identify pioneer species that colonize Fe tailings naturally and concentrate heavy metals. From the second year on, nine species of herbaceous species, four species of tree including *Alstonia scholaris*, *Azadirachta indica*, *Peltaphorum* and *Tectona grandis* were found to grow naturally. It was observed that certain species accumulatea high concentration of metals, illustrating the internal detoxification of heavy metals.

6 Radionuclide Uptake in Selected Plant Species for Consolidation

Ecological disturbance occurs in mine areas due to interference in the plant, soil, geological, hydrological and social fabric of the ecosystem. Minerals are formed over a period of centuries through the process of fossilization of biological materials and their conservation into different mineral strata. During the process of mining, these natural resources are exploited to obtain valuable minerals for development of the society. As a result of mining, these natural resources are removed from the body of the earth, processed and used in industries to keep pace with the developmental cycle. It is therefore expected that the process of industrialization, including mining, may lead to ecological disturbance, if control measures are not exercised. However, the degree of disturbance depends on the mode of mining, that is, underground or surface, the location of the mining site and the ecological, hydrological, geological and sociological conditions of the mined areas.

Disposal of mine tailing wastes by landfill in the form of slurry is the most widely practiced method in the world, including in India. Therefore, radionuclide and metal pollution is a global environmental problem, and the number of contaminants entering the environment has increased greatly in recent times due to increased mining activities (Usman and Nasiru, 2011). Long-term stabilization of radionuclides and other pollutants from contaminated substrates is a key criterion for the success of rehabilitation works. *In situ* soil remediation at sites with low levels of U contamination, involving the use of chemical stabilizing agents and subsequent revegetation, is attractive because it is expected to be more cost effective and less disruptive to the environment.

The movement and subsequent off-site dispersal of radionuclides are essential factors for risk assessments and determining remedial actions at hazardous, radioactive waste sites. One potential transport path from the soil column is through the uptake and deposition of plant material to surface soil. The behavior and mobility of radionuclides in soil is a major consideration for plant uptake and is influenced by many variables (Romney et al., 1974; Poston et al., 1995b). The uptake characteristics of radionuclides have been observed to vary depending on soil properties such as texture, organic matter content, bacterial action, pH, redox potential and physiochemical speciation (Poston et al., 1995; White and Dunaway, 1975; Romney et al., 1974; Han and Lee, 1997; Salbu and Oughton, 1995).

The two most significant radionuclide that may appear in the environmental matrices are U (nat) and ^{226}Ra in view of the waste fixation technology. The former is chemically more toxic and accumulates in the kidney, whereas the latter is a bone seeker with behavior similar to calcium (a known essential element for bone

DOI: 10.1201/9781003326151-6

development). Consequently, from the point of view of abundance, established chemical/radiological toxicity and long half-lives, these two elements are the most significant radionuclides in the uranium processing industry. In the text that follows, the terms "uranium" and "radium" refer to natural uranium and ^{226}Ra unless mentioned otherwise.

Tracy et al., (1983) and Popova et al., (1964) found radium to be localized predominantly in the roots of the plants but also noted elevated concentrations in leafy vegetables such as spinach and lettuce. They hypothesized that radium is translocated to areas of maximum plant growth, such as root tips and terminal leaves, and then becomes permanently fixed in these plant parts. Mistry et al., (1965) also noted the highest concentration of ^{228}Ra in leaves. D'Souza and Mistry (1970) showed ^{226}Ra to be transferred rapidly from root to shoot in beans plants. Linsalata et al. (1989) are of the view that the correlation existing in the concentration ratio values between elements with diverse chemical properties (viz., Ra, Th) may be linked to the metabolic requirement of the plant species.

In most of the literature reviewed, it was found that radium, in general, has been reported to decrease within various plants as roots >stem and leaves > seed/fruits. D'Souza and Mistry (1971) reported that the distribution pattern of radium was similar to that of calcium within different tissues of the same plant. According to Bollard (1960), calcium is regarded as the most immobile of the essential elements in plants, exhibiting little translocation once delivered to a particular organ.

Current remediation technologies fall into four basic categories, namely physical, chemical, biological and thermal. Thermal treatment will remove contaminants but will destroy other soil characteristics beneficial for plant growth. Chemical extraction and soil washing can also degrade the quality of the soil. To study the restoration of contaminated soils or degraded land, the effect of amending a land soil and to bioremediated soils with green waste compost and their subsequent ability to support the growth of grass and trees may provide a sustainable solution. Phytoremediation is a natural process carried out by plants that are able to live in a contaminated media. Hyperaccumulators are plants that can absorb high levels of contaminants with their roots and concentrate them either within their roots, shoots, and leaves. A variety of plant species are commonly used to remove heavy metal from the soil. The use of soil amendments can be a cost-effective in situ process for remediation, revitalization and reuse.

According to Jha et al. (2001) concentration of uranium in the roots of Typha latifolia (cat-tail) was found to vary within narrow range i.e. of the same order (57–222.5 Bq/kg) with geometric mean and arithmetic mean 117 and 135 and geometric standard deviation and standard deviation 44.25 and 70.25 Bq/kg respectively. Shoot of the same species was showing wide variation and some order of difference was found in its shoot, ranging from 0.05 to 22.75 Bq/kg with geometric mean and arithmetic mean 1.5 and 6.35 and geometric standard deviation and standard deviation 260.25 and 8.0 Bq/kg. It appears that the transfer of uranium is a gradual process depending on the maturity status of the plant and plant on its mature stage retains the accumulated uranium in the shoot. However, as reported elsewhere, fruits and seeds uranium content is least as compared to root, stem or leaf. Retention capacity of shoot appears to be poor and is evident from the analysis of other species as well.

Tracy et al. (1983) and Popova et al. (1964) found radium to be localized predominantly in the roots of the plants but also noted elevated concentrations in leafy vegetables such as spinach and lettuce. They hypothesized that radium is translocated to areas of maximum plant growth, such as root tips and terminal leaves, and then becomes permanently fixed in these plant parts. Mistry et al. (1965) also noted the highest concentration of [228]Ra in leaves. D'Souza and Mistry (1970) showed [226]Ra to be transferred rapidly from root to shoot in beans plants. Linsalata et al. (1989) are of the view that the correlation existing in the concentration ratio values between elements with diverse chemical properties (*viz.*, Ra, Th) may be linked to the metabolic requirement of the plant species.

Radium uptake by plants growing over buried uranium mill tailings could still be of concern. Similar problems of dispersal are found with some mine spoils from uranium mines, with storage of uranium ore and with phosphate mining. Thus, radium can be enhanced in the environment by these activities. This study provides the basis for estimation of intake level by herbivores and for evaluating transfer coefficients. Additional information for a larger number of species and for different tailings materials and soils could be useful in making these estimates. The study showed the bioavailability of radium for plant uptake by several native and neutralized species growing at diverse uranium mill sites throughout the Four Corners region of the southwestern United States (Marple and Potter, 1982).

According to Jha et al. (2001), the highest concentration among all the species analyzed was found in the different organs of the *Ipomoea aquatica* (family Convolvulaceae). Concentration in root, stem and branch and leaf was 4044, 464 and 442 Bqkg[-1]. Concentration in stem and branch and leaf was comparable to the root of other species like *Typha latifolia* and *Saccharum spontaneum* growing over a tailing pile. The plant may thus be taken as a hyperaccumulator of [226]Ra.

Knowledge of the levels of the naturally occurring radionuclides is important because they contribute a substantial portion of the radiation dose to humans, and diet is usually the major source of these radionuclides (NCRP, 1975; UNSCEAR, 2000). Polonium-210 (T1/2 = 138 days), an alpha emitter of the uranium-238 series, is of radiological interest for a number of reasons, but mainly because of its large contribution to the internal radiation dose (Skwarzek, 2000). Evidence from the literature indicates that polonium-210 is accumulated strongly by organisms and transferred via food along the trophic chain (Heyraud and Cherry, 1979; Skwarzek and Falkowski, 1988; Khan et al., 2002). Because of the possibility that large amounts of 210Po may be thus transferred to humans via diet, there is a need for the investigation of this nuclide in foodstuffs.

Uranium uptake studies of various plant species in different regions of the world have been carried out by different workers (Table 6.1). Uptake of uranium was believed to be in the anionic complex form. Cannon, 1960), in a review of literature from botanical prospecting of uranium on the Colorado plateau, mentioned that when either sulfur or selenium is added, there is increase in solubility of uranium in the carnotite plot. This was believed to be a result of oxidation to uranyl sulfate and uranyl selenate in the soil solution.

When gypsum and selenium were added to the carnotite plot, uranium was found to be absorbed easily and transported to the upper part of the plant. When phosphate and lime were added, absorption of uranium was found to be lessened. Results of laboratory experiments with soil constituents imply that the availability of uranium may be increased in a carnotite environment due to the formation of uranyl sulphate (Cannon, 1960). Generally, more uranium is found in root than in the parts of the plant above ground, although the ratio of these metals in the roots to that of the tops show considerable variation. If the amount precipitated in the root is large, the root cell may become clogged and the plant may die. A study of the uranium content of juniper trees and their roots was made in the Thompson district of Utah. It was found that the elements tend to precipitate near the point of intake in the root, lesser amounts are found in the root approaching the ground surface and even smaller amounts are found in the limbs and branches. Forty near-surface juniper roots contained five to six times as much uranium as the branch tips. Moreover, two samples of juniper root at depth contained 70 and 200times as much uranium as the branch tips (Cannon, 1960). Additional information was obtained from analysis of entire root systems of the plants grown in the experimental garden. Plant species that are known to accumulate large amounts of uranium in the foliage contained less uranium in the roots than species that do not accumulate these elements in the aboveground part of the plant. The variation of a particular metal in a plant, however, may not accurately reflect its content in the soil, but may instead be related more closely to variations in absorption of some other ions. Plants that absorb the most uranium on the Colorado plateau are sulphur and selenium indicator plants. Yet the concentration of uranium in these plants is not closely allied to uranium content of the soil but to the degree of oxidation and to the selenium and sulphur content of the regolith. The uranium content of these plants cannot be used as a reliable indication of mineralized grounds. Under oxidizing conditions, the uranyl ion (UO_2^{++}) is the most likely form of uranium that can be present the environment. Uranium accumulates in plants, and its mobility in the environment is affected by microorganisms (Cannon, 1960).

Sheppard and Evenden (1983) briefly discussed the uptake of uranium by plants under two different physicochemical environments—sand and peat soil under laboratory conditions at the same level of induced uranium. The inference of the said work indicates that in the case of uranium, plant mediation plays a more important role, since essentially no uranium translocation occurs in plants. Uranium may be excluded from the plant by an active root membrane selection process. Uranium uptake by Swiss chard from sand is up to 80 times higher than from peat at the same level of applied uranium. However, as previous studies suggest, it is restricted to the root system and may only be precipitated on the outer root membrane and may not accumulate in the root.

Miera et al. (1980), while discussing the mobility of elevated levels of uranium in the environment, have discussed the difficulty of interpreting the analytical results of root concentration. Whether the sorbed uranium is penetrated inside the root or merely present at the outer surface is not clear. Variability in root/soil uranium concentration ratio observed was probably due to the presence of small soil particles (observed by microscopic examination), presumably uranium adhering to the roots. It can be further speculated that uranium colloids may have been sorbed on the root surface. This further complicates the differentiations of "in" vs. "on" uranium components in plant roots. Ebbs et al. (1998), through an uptake study by hydroponic techniques, concluded that at pH 5.0, the uranyl (UO_2^{++}) cation was more readily taken up and translocated by peas (*Pisum sativum*) than the hydroxyl and carbonate U complexes present in the solution at pH 6.0 and 8.0, respectively. A subsequent experiment tested the extent to which various monocot and dicot species take up and translocate the uranyl cation. Of the species, screened tepary bean (*Phaseolus acutifolius*) and red beet (*Beta vulgaris*) showed the greatest accumulation of uranium. The initial characterization of uranium uptake by peas suggested that in the field, a soil pH of <5.5 would be required in order to provide uranium in the most plant available form. A pot study using uranium-contaminated soil was therefore conducted to access the extent to which two soil amendments of HEDTA and citric acid, were capable of acidifying the soil, increasing U solubility, and enhancing U uptake by red beet. Of these two amendments, only citric acid proved effective, decreasing the soil pH to 5.0 and increasing uranium accumulation by a factor of 14. The result of this pot study provides a basis for the development of an effective phytoremediation strategy of uranium-contaminated soil. It was found that only the uranyl ion, which has a plus two charge and forms in soil solution under acidic conditions, accumulated in plant tops.

A comparative uptake study of uranium and radium was carried out by Rumble and Bjugstad (1986) at the uranium mill tailings disposal site and control areas of South Dakota. A higher concentration of uranium was observed in plants growing on uranium mill tailings than in those from the control sites. Rickard et al., (1977) suggested that plant uptake of uranium is independent of soil concentration. It was reported by the same author that the concentration of uranium in soil is greater than 200 ppm. These are toxic to some plants, and uptake of uranium from the soil by plants is in the range $10^{-1}10^{-2}\mu g.g^{-1}$ calculated on a fresh weight basis. The higher concentration ratios for uranium on the mill tailings would suggest that there is some dependence of uptake of these radionuclides on initial soil concentration. The long-term implication of radionuclide uptake involves the accumulation of uranium on the soil surface from many years of litter fall and decomposition, as well as off-site movement by wind and water.

Morten et al., (2002) has given a detailed discussion on dependence of plant organ concentration and concentration ratio on soil parameters (pH, delta pH, total soil concentration, etc.). Further, the interrelationship between tissue concentrations (stem and leaf) was also discussed by the author. Two horizons of soil, that is AE and Bs, were selected for evaluation of concentration ratio; apart from this, the long-term effect of plant uptake on recycling was also presented in the review. These authors studied the extent of uranium and thorium uptake and recycling by blueberry (*Vassinium pallidum*+ Aiton) in native habitat and identification of the soil properties and processes that contribute most to uranium and thorium bioavailability in this system. Both AE and Bs horizon soil concentrations were determined and the concentration ratio for stems and leaves were evaluated based on the soil concentration of both horizon and tissue (stems and leaves) concentrations. Mean uranium concentration was 8.65×10^{-3} $\mu g.g^{-1}$ dry in leaf and 7.95×10^{-3} $\mu g.g^{-1}$ dry in stem. Uranium concentration in leaves shows little preferential translocation. Moreover, uranium uptake was found to be sensitive with respect to soil properties. Plant tissue concentrations of radionuclides have rarely shown a linear relationship to radionuclide concentration in substrate (Sheppard and Sheppard, 1985; Sheppard and Evenden, 1988; Sheppard et al., 1989). Plant and soil concentrations were not significantly

TABLE 6.1

Uranium Uptake Studies in Different Geographical Regions (after Singh et al. 2021)

Uranium uptake	Plant species	Uptake range (Bq/kg)	References
Uranium mine tailing pond-2, Jharkhand state, India	*Colebrookeaoppositifolia, Dodonaea viscosa, Furcraea foetida, Imperata cylindrica, Jatropha gossypifolia, Pogostemon benghalense* and *Saccharum spontaneum*	10–28	Singh and Soni (2010a,b)
Uranium mine tailing pond-1, Jharkhand state, India	*Azadirachta indica, Bacopa monnieri, Croton bonplandianus, Calotropis procera, Penstemondigitalis, Pteris vittata* and *Saccharum spontaneum*	64–155	Singh et al. (2015)
Neuensalz-Zobes, Western Ore Mountains, Saxony	*Phragmites australis*	24–87	Wang et al. (2015)
Uranium mill tailing pond-3 at Jaduguda, India	*Pistia* sp., *Eichhornia crassipes, Jussiaea* sp., *Hydrilla verticillata, Vallisneria* sp., *Polygonum barbatum, Nymphaea* sp., *Marsilea* sp.	0.002–29.03	Jha et al. (2016)
Uranium and thorium in the tailing water of Keban, Turkey	*Lemna minor* and *Lemna gibba*	22–26.5	Sasmaz et al. (2016)
Uranium-contaminated Sevilha mine, Central Portugal	*Juncus squarrosus, Carlina corymbosa, Juncus bufonius, Callitriche stagnalis, Lemna minor, Riccia fluitans*	285–742.5	Favas et al. (2016)
Uranium mill tailings, hamlet in Slovenia, Upper Carniola region of Slovenia. Pot experiment	*Raphanus sativus* L., *Brassica oleracea var. sabauda, Diplotaxis tenuifolia*	19–895	Planinsek et al. (2016)
Uranium mill tailings, southern part of China	*Cyperus iria* and *Miscanthus floridulus*	0.13–22.1	Yan (2016)
Uranium mill tailings, northeast part of China	*Artemisia mongolica* and *Oenothera biennis*	145.2–525	Yan et al. (2018)

TABLE 6.1

Uranium uptake	Plant species	Uptake range (Bq/kg)	References
Tummalapalle uranium mining area, Cuddapah district, Andhra Pradesh, South India	*Rivea ornata, Tribulus terrestris, Pergularia daemia, Allmania longipedunculata, Albizia amara, Ricinus communis, Scleria lithosperma, Flugiamonophylla, Cymbopogon citratus, Calotropis gigantea*	53.28–123.51	Sreedhar et al. (2017)
Uranium mill tailings, hamlet in Slovenia, Upper Carniola region of Slovenia. Pot experiment	*Brassica rapa* L. subsp. *pekinensis* (Lour.) *Hanelt*	0.6–4.7	Cerne et al. (2018)

correlated; the correlation coefficient between AE horizon soil and leaf concentration was −0.13, whereas it was −0.41 for stem. Plots of CRs identify a clear differentiation by tissue type and by nuclide. Although CRs based on AE horizons are somewhat higher, CR trends were virtually identical for both AE and Bs horizons. CRU_{lv} values were not significantly different from CRUst. When CRs were calculated based on AE horizons, CRTh in both stem and leaves declined with increase in soil concentration. Additionally $CRTh_{AE}lv$ and $CRTh_{AE}st$ were highly correlated (R = 0.88, p = 0.001), but $CRU_{AE}lv$ values were poorly related to $CRU_{AE}st$. Studies by Sheppard and Evenden (1988) found that U and Th CR values decrease as soil concentration increases. Control pot study of blueberry plants grown in organic soil artificially contaminated with uranium were showing CR values nearly an order of magnitude higher as reported by Morten et al. (2002), although CR values decline significantly with time. This suggests plant uptake may eventually equilibrate. Neither uranium plant concentration nor uranium CRs were linearly related to uranium soil concentrations suggesting that soil uranium may be partitioned more explicitly into available and unavailable forms. Weathering, processes responsible for particle size reduction and increased surface area appear to enhance the release and availability of uranium. Studies carried out at the Anaconda mill site for forage uptake of uranium series radionuclides has given an uptake coefficient for ^{238}U (ratio of vegetation activity to soil activity) that varies from 0.69 in Indian rice grass roots to 0.01 in four-wing saltbush (*Atriplex canescans*) shoots (Rayno et al.,1980).

Uranium concentrations in some crops (high-yielding *Brassica* family, root crops such as beets, Swiss chard and sugar beet and high biomass crops like corn, sunflower, *Cynodon dactylon*, etc.) were evaluated by application of uranyl nitrate solution and EDTA complexing agent in the ratio (1:1). Plants were harvested after six to eight weeks, and the uranium in the dry matter content of various plant organs were determined. Generally, addition of EDTA increased uranium concentration in the roots and shoots of the *Brassica* species 10% to 35%. For some plant species, the addition of EDTA has little effect on uranium accumulation. Root crops and high biomass crops differed in uranium uptake and accumulation. Some plants absorbed and translocated uranium to a greater extent than did others. Among the plant species tested, sunflower accumulated the most uranium, mainly in the root (800 μg/g dry). Sugar beet had the highest uranium concentration in the shoots and Bermuda grass and corn had the lowest. Except for Swiss chard and corn, the addition of EDTA significantly increased uranium concentration in shoots of all plants (Hossner et al.,1997).

Jovanovic et al. (2004) has studied uptake and distribution of uranyl nitrate in soybean (*Glycine max*), sunflower (*Helianthus annus*) and maize (*Zea mays*) plants. The effect of uranyl nitrate application on plant physiology was also studied by the author for a definite follow-up period (12 hr–100 hr) after application. No serious damage for the plants was observed for the entire determination period after application of 150 and 300 ppm uranyl nitrate solution. However, some slight leaf damage was visible on sunflower plants after 100 hours. Uranium content in roots rapidly increased after 12 hours, nearly doubling after 100 hours in sunflower and soybean. On the other hand, the content of uranium in maize roots after initially increasing 20% continued to constantly decrease until the end of the experiment. The average uranium content in the entire plant was highest for sunflower, lower for maize and lowest for soybean (5056, 2264 and 1835 respectively in relative units). Uptake rate calculated from the first part of the uptake curve was the highest in sunflower, lower in soybean and the lowest in maize. To conclude, sunflower was found to be highly suitable for uranium remediation of nutrient solution, as its uranium uptake is fast and the amount taken up is high.

Gunther et al. (2003) studied the uranium speciation after uptake of uranium by lupine plants. Differences were detected between the uranium speciation in the initial solution (hydroponic solution and pore water of soil) and inside the lupine plants. The oxidation state of uranium did not change and remained hexavalent after it was taken up by the lupine plants. The chemical speciation of uranium was identical in the roots, shoot axis and leaves and was independent of the uranium speciation in the uptake solution. The results indicate that the uranium is predominantly bound as uranyl (VI) phosphate to the phosphonyl groups.

Singh and Soni (2010a) worked on seven species that were effective soil binders and suitable soil conserving species for tailing pond areas. All these species were good soil binders. Vasistha (1992) studied the root penetration studies and soil binding capacity of primary colonizing species on rock phosphate mines of Doon and concluded that *Buddleja asiatica*, *Wendlandia exserta* and *Trema politoria* are effective soil binders and suitable soil-conserving species on degraded mined lands.

6.1 NATURAL URANIUM, POLONIUM AND RADIUM IN SOIL AND TAILINGS

Natural uranium occurs in uranium deposits, which are mined and processed into nuclear materials for military and civilian uses. The modern uranium mining industry began in the 1940s. As the demand for nuclear fuel has increased, uranium mining and milling activities have been extensively conducted in the past decade (Carvalho et al., 2014; Smodiš et al., 2015). The effective dose rate produced by radionuclides that were released from worldwide mining, hydrometallurgy and tailings from 1998 to 2002 was approximately 0.238 Gy^{y-1} (United Nations Scientific Committee on the Effects of Atomic Radiation, 2010). The soil's affinity for uranium and other soil characteristics, such as its acidity or alkalinity and water content, affect the mobility of weathered uranium in the soil profile. Water should therefore be low in soils that have a high affinity for uranium binding and release (usually soils high in organic matter). Accordingly, mobility is probably larger in soils that bind uranium less tightly, such as semiarid climatic soils with neutral to alkaline soil pH and low organic carbon content (Ebbs et al., 1998). The concentration of uranium in the soil had the biggest impact on plant uranium levels.

It is well known that mining of uranium ore produces large amounts of tailings that contain high concentrations of uranium-series radionuclides. Long-lived radionuclides of the ^{238}U decay series, specifically ^{230}Th, ^{226}Ra, ^{210}Pb, and ^{210}Po, are the main source of concern with relation to environmental security. 80,000 years, 1622 years, 22 years, and 138 days, respectively, are their half-lives. Due to the low concentration of ^{235}U in the ore (0.7%), the environmental impact of its daughter products is less significant..

Uranium has two geologically significant oxidation states: +6 and +4. In the oxidized state, uranium is quite mobile. The uranyl ion UO_2^{++} forms bi- and tri carbonate complexes that move easily through the oxidized environments. A phosphate complex may also be an important source of uranium mobility. During weathering, the uranium minerals yield hydrated uranium oxides, which are sparingly soluble. Some uranium remains in the highly resistant minerals and may become concentrated in placers. Only a small part of uranium goes into solution, preferably as the readily soluble uranyl complexes, which are understood as basic salts containing UO_3. The soluble uranium compounds are readily adsorbed on the hydroxide gel of Fe, Al and Mn and silica gel. Uranium may change its ionic charge comparatively readily according to the circumstances prevailing in the surroundings. The transition from +4 to +6 has a redox potential within the normal range of geological environments.

Most plants take up their nutrients and contaminants mainly through the roots however, there is also absorption by leaves, from the soil solution. The quantity of soluble soil uranium that is available for uptake by plants depends on how strongly and to what extent it is bonded to soil components. Therefore, it is expected that the factors affecting uranium mobility in soil will have a significant impact on how contaminated plants are. The uptake of uranium by plants is larger than that of plutonium and americium, although being less than that of mobile radioactive elements like cesium and strontium. The soluble forms of uranium seem to be readily absorbed by the plants. However, in many soils, natural uranium has a low solubility and can be unevenly distributed. In general, uranium concentration in plants decline in the order: root >shoot > fruit and seeds. A stimulatory effect on growth has been observed in some grasses exposed to elevated concentrations of uranium in soil at broad concentrations. The concentration ratio has been determined for different matrices: mine waste, tailings, processed waste and so on (Royal Society, USA, 2000). It has been pointed out that the concentration ratio varies by four orders of magnitude for crops on different soils and different soil uranium. This wide variation severely inhibits the applicability of generic models that incorporate uptake by plants and highlights the need-based approach with well-defined source terms and composition (Royal Society, 2000).

However, there is also absorption by leaves, from the soil solution. The quantity of soluble soil uranium that is available for uptake by plants depends on how strongly and to what extent it is bonded to soil components. Therefore, it is expected that the factors affecting uranium mobility in soil will have a significant impact on how contaminated plants are. The uptake of uranium by plants is larger than that of plutonium and americium, although being less than that of mobile radioactive elements like cesium and strontium.

It was thought that uranium was absorbed as an anionic compound. When either sulphur or selenium is added, there is an increase in uranium's solubility in the carnotite plot, according to Cannon (1960), who reviewed the literature from botanical prospecting for uranium on the Colorado plateau. This was thought to be the result of oxidation in the soil solution to uranyl sulphate and uranyl selenate.

When gypsum and selenium were added to the carnotite plot, uranium was found to be absorbed easily and transported to the upper part of the plant. When phosphate and lime were added, absorption of uranium was found to be lessened. Results of laboratory experiments with soil constituents imply that the availability of uranium may be increased in a carnotite environment due to the formation of uranyl sulphate (Cannon, 1960). Generally, more uranium is found in root than in the parts of the plant above ground, although the ratio of these metals in the roots to that of the tops show considerable variation. If the amount precipitated in the root is large, the root cell may become clogged, and the plant may die. A study of the uranium content of juniper trees and their roots was made in the Thompson district of Utah. It was found that the elements tend to precipitate near the point of intake in the root, lesser amounts are found in the root approaching the ground surface and even smaller amounts are found in the limbs and branches. Forty near-surface juniper roots contained five to six times as much uranium as the branch tips. Moreover, two samples of juniper root at depth contained 70 and 200 times as much uranium as the branch tips (Cannon, 1960). Additional information was obtained from analysis of entire root systems of the plants grown in the experimental garden. Plant species that are known to accumulate large amounts of uranium in the foliage contained less uranium in the roots than species that do not accumulate these elements in the aboveground part of the plant. The variation of a particular metal in a plant, however, may not accurately reflect its content in the soil, but may instead be related more closely to variations in absorption of some other ions. Plants that absorb the most uranium on the Colorado plateau are sulphur and selenium indicator plants. Yet the concentration of uranium in these plants is not closely allied to uranium content of the soil but to the degree of oxidation and to the selenium and sulphur content of the regolith. The uranium content of these plants cannot be used as a reliable indication of mineralized grounds. Under oxidizing conditions, the uranyl ion (UO_2^{++}) is the most likely form of uranium that can be present the environment. Uranium accumulates in plants, and its mobility in the environment is affected by microorganisms (Cannon, 1960).

The two most significant radionuclides that may appear in the environmental matrices are U (nat) and ^{226}Ra in view of the waste fixation technology. The former is chemically more toxic and accumulate sin the kidney, whereas the latter is a bone seeker with behavior similar to calcium (a known essential element for bone development). Consequently, from the point of view of abundance, established chemical/radiological toxicity and long half-lives, these two elements are also the most significant radionuclides in the uranium processing industry.

Some uranium emissions take place during the processing and extraction of uranium ore. Radium emissions and other uranium decay products are typically more significant than uranium in tailing ponds at mines and ore processing facilities (Wikipedia, 2012).

A comparative uptake study of uranium and radium was carried out by Rumble and Bjugstad (1986) at the uranium mill tailings disposal site and control areas of South Dakota. A higher concentration of uranium was observed in plants growing on uranium mill tailings than in those from the control sites. Rickard and Keough (1968) suggested that plant uptake of uranium is independent of soil concentration. It was reported by the same author that the concentration of uranium in soil is greater than 200 ppm. These are toxic to some plants, and uptake of uranium from the soil by plants is in the range $10\text{-}^1\text{-}10\text{-}^2\mu g.g^{-1}$ calculated on a fresh weight basis. The higher concentration ratios for uranium on the mill tailings would suggest that there is some dependence of uptake of these radionuclides on initial soil concentration. The long-term implication of radionuclide uptake involves the accumulation of uranium on the soil surface from many years of litter fall and decomposition, as well as off-site movement by wind and water.

Morten et al. (2002) has given a detailed discussion on dependence of plant organ concentration and concentration ratio on soil parameters (pH, delta pH, total soil concentration, etc.). Further, the interrelationship between tissue concentrations (stem and leaf) was also discussed by the author. Two horizons of soil, that is AE and Bs, were selected for evaluation of concentration ratio; apart from this, the long-term effect of plant uptake on recycling was also presented in the review. These authors studied the extent of uranium and thorium uptake and recycling by blueberry (*Vassinium pallidum*+ Aiton) in native habitat and identification of the soil properties and processes that contribute most to uranium and thorium bioavailability in this system. Both AE and Bs horizon soil concentrations were determined and the concentration ratio for stems and leaves were evaluated based on the soil concentration of both horizon and tissue (stems and leaves) concentrations. Mean uranium concentration was 8.65×10^{-3} µg g^{-1} dry in leaf and 7.95×10^{-3} µg g^{-1} dry in stem. Uranium concentration in leaves shows little preferential translocation. Moreover uranium uptake was found to be sensitive with respect to soil properties. Plant tissue concentrations of radionuclides have rarely shown a linear relationship to radionuclide concentration in substrate (Sheppard and Sheppard, 1985; Sheppard and Evenden, 1983; Sheppard et al., 1988). Plant and soil concentrations were not significantly correlated; the correlation coefficient between AE horizon soil and leaf concentration was −0.13, whereas it was −0.41 for stem. Plots of CRs identify a clear differentiation by tissue type and by nuclide. Although CRs based on AE horizons are somewhat higher, CR trends were virtually identical for both AE and Bs horizons. CRU$_{lv}$ values were not significantly different from CRUst. When CRs were calculated based on AE horizons, CRTh in both stem and leaves declined with increase in soil concentration. Additionally $CRTh_{AE}lv$ and $CRTh_{AE}st$ were highly correlated (R = 0.88, p = 0.001), but $CRU_{AE}lv$ values were poorly related to $CRU_{AE}st$. Studies by Sheppard and Evenden (1988) found that U and Th CR values decrease as soil concentration increases. Control pot study of blueberry plants grown in organic soil artificially contaminated with uranium were showing CR values nearly an order of magnitude higher as reported by (Morten et al., 2002). Although CR values decline significantly with time. This suggests plant uptake may eventually equilibrate. Neither uranium plant concentration nor uranium CRs were linearly related to uranium soil concentrations suggesting that soil uranium may be partitioned more explicitly into available and unavailable forms. Weathering, processes responsible for particle size reduction and increased surface area appear to enhance the release and availability of uranium. Studies carried out at the Anaconda mill site for forage uptake of uranium series radionuclides has given an uptake coefficient for ^{238}U (ratio of vegetation activity to soil activity) that varies from 0.69 in Indian rice grass roots to 0.01 in four-wing saltbush (*Atriplex canescans*) shoots (Rayno et al.,1980).

Rufykiri et al., (2001) discussed that arbuscular mycorrhizal fungi can transport uranium under root organ culture conditions. An experiment related to the transport of uranium by arbuscular mycorrhizal fungi showed that uranium was mobilized by AMF. Mycorrhizal carrot roots exposed to a low level of uranium mobilized more uranium per biomass unit than did nonmycorrhizal carrot roots. This AMF effect was not observed at relatively high uranium levels. Uranyl cation (UO_2^{2+}) is stable under acidic conditions, phosphate and hydroxyl complexes under acidic or near neutral conditions, while carbonate complexes predominate under alkaline conditions. Noteworthy was the high uranium transport to roots observed at pH 4, indicating that soluble uranyl cation or uranyl-sulphate species were more mobilized by fungal tissues, while phosphate species dominating at pH 5.5 or hydroxyl and carbonate species dominating at pH 8 were rather immobilized by hyphal structures. The high bioavailability of uranyl cation was also reported in another study involving plants (Ebbs et al., 1998). Such results suggest that pH is an important factor whose influence cannot be ignored when studying the mechanism of uranium bioavailability.

Hass et al., (1998) studied bioaccumulation of uranium by lichens. The uptake of aqueous uranium by *Peltigera membranacea* was found to be a function of time and pH. It was determined that uranium sorption was strongest in the pH range 4–5 with maximum sorption occurring at pH of 4.5 and an incubation time of 24 hours. Similarly, experiment and field studies by Mann and Fyfe (1995); Milodowski et al. (1990); Lovey et al., (1991) and Macaskie et al., (1992) demonstrated that algae and bacteria can facilitate the precipitation of solid uranium phases directly from solution, through absorption of aqueous U into bacteria cell walls and subsequent nucleation of hydrous uranyl phases (Macaskie et al., 1992) or through uranium absorption followed by reduction of UO^{2+} to relatively insoluble U^{4+} at the cell wall interface (Lovey et al., 1991). Algae may bioaccumulate uranium through a combination of mechanisms including surface complexation, biomineralization and physical trapping of dust and soil particles.

Jovanovic et al., (2004) has studied uptake and distribution of uranyl nitrate in soybean (*Glycine max*), sunflower (*Helianthus annus*) and maize (*Zea mays*) plants. The effect of uranyl nitrate application on plant physiology was also studied by the author for a definite follow-up period (12 hr–100 hr) after application. No serious damage for the plants was observed for the entire determination period after application of 150 and 300 ppm uranyl nitrate solution. However, some slight leaf damage was visible on sunflower plants after 100hours. Uranium content in roots rapidly increased after 12hours, nearly doubling after 100hoursin sunflower and soybean. On the other hand, the content of uranium in maize roots after initially increasing 20% continued to constantly decrease until the end of the experiment. The average uranium content in the entire plant was highest for sunflower, lower for maize and lowest for soybean (5056, 2264 and 1835 respectively in relative units). Uptake rate calculated from the first part of the uptake curve was the highest in sunflower, lower in soybean and the lowest in maize. To conclude, sunflower was found to be highly suitable for uranium remediation of nutrient solution, as its uranium uptake is fast and the amount taken up is high.

Uranium speciation following uranium uptake by lupine plant was examined by Gunther et al., in 2003. There were variations in the uranium speciation between the original solution (hydroponic solution and soil pore water) and the lupine plants.

After being absorbed by the lupine plants, uranium's oxidation state did not change and remained hexavalent. The chemical speciation of uranium in the roots, shoot axis, and leaves was the same and distinct from the chemical speciation of uranium in the uptake solution. According to the findings, the uranium is primarily linked to the phosphonyl groups as uranyl (VI) phosphate.

Clay and organic matter are active ingredients for nourishing plants, according to soil science literature. Because they have negative surface charges, they have a high cationic adsorptive capacity. The pH, moisture content, cation-exchange capacity, and competing ions are among the factors that can influence the migration of radionuclides and trace elements from soil/substrate (NCRP, 1992). In addition, factors like soil texture and temperature in the environment play a role in determining the degree of bioaccumulation in plants. Before being connected to a metabolic system, radionuclides and trace elements travel over a complicated journey.

Several researchers have worked on the impoundment for the tailings from uranium mills.Markose (1990) evaluated the physical properties of the tailings' (particle size) leach ability and related radioactivies. There were also prior attempts.(Webb and Voorhees, 1984) to assess how tailing piles are dispersed due to various natural processes and engineering blunders. Eisenbud (1987) presented the techniques for lowering radon emission by covering the tailings with dirt. The migratory properties of radionuclides from the tailing pile were evaluated by Gans et al., in 1981. The nature and spread of pollutants from the uranium ore processing sector were reviewed by Tsivoglue and O'Connel in 1963. Morishima et al., (1976) studied the movement of uranium in soil and vegetables.

Cannon(1952)reviewed uranium uptake by plants grown over uranoum and vanadium deposits in Colorado plateau.Because they contain the entire spectrum of radionuclides from the ^{238}U decay series, uranium mill tailings are the main source of

^{226}Ra. As a result, the migration and accumulation of "radium" in the abiotic and biotic components of the adjacent environment represent a critical link in the disposal of processed waste from the uranium industry. Because of radium's high level of radiotoxicity, long radiological half-life, and metabolic connection to calcium (a crucial nutrient in humans), evaluation of its migratory properties is of concern. Additionally, a number of workers have reported radium being released into a nearby environment close to a uranium company.

6.2 ROOT SHOOT GROWTH OF SELECTED PLANTS

A clear picture of root shoot growth of plants useful for restoration of mining/tailing sites is an important aspect of the effectiveness of species for sustainably stabilizing and remediating affected areas. Different workers have reported studies on plant form and function (Robson et al., 2015) and morphogenetic aspects of plant growth (Chapman and Lemaire, 1993) and root form and function (Dawson et al., 2000).

It may be observed that the stability of soils during the early phases of development is influenced by the belowground competition and the enhanced responsiveness of roots on a restored site (Nambier, 1981). The total root length measurements made at both locations are well within the parameters set by Carbon et al. (1980) and Nambier (1983). Shrubs should occupy a significant portion of the soil volume given their extensive root system, making their significance in soil physics clear. A mine site's dry, infertile soils have undergone a significant edaphic alteration as a result of this extensive root network.

Several researchers have noted improvements in soil properties brought about by the addition of organic matter through root degradation (Thakur, 1957; Bhimaya and Seri, 1980; Singh and Gupta, 1983).

In general, the longer roots in soils from natural habitats may not be as significant as their value for plants cultivated in ecosystems that have been degraded, where soil moisture can be seen as the primary factor limiting plant survival and growth. The aforementioned species have an advantage in that they can absorb more water since they have more lateral roots and larger root surfaces. These species have a stronger capacity for growth across stress areas because of this characteristic. It can be hypothesised that these species with a proportion of branching roots would absorb more minerals per unit of weight than would unbranched roots as nutrient absorption also depends on root length and surface area (Nye, 1966).

Case study (Chapter 7) reveals that the comparative root length of plant species on a revegetated tailing pond site (*in situ*) and experimental containers site (*ex situ*) that the length of roots varied in both species, as well as between the two contrasting sites. *Pogostemon benghalense* proved to be superior in terms of its contribution to total root length on revegetated site as well as natural habitat. The extensive growth of this species was undoubtedly due to the presence of an extensive network of fibrous roots. The measurement of root system shows that the major portion of the roots was found in the upper 25 cm of soil. The data are in agreement with findings from other studies (Schafer et al., 1980; Roberts et al., 1988).This led to an increase in organic carbon content in the upper surface layer of restored mine soil. The increase in organic carbon in restored area may be ascribed to the addition of plant litter and other residues to the soil surface.

While there are many benefits to vegetation and plant growth over treated tailings, there are also some drawbacks. Vegetation's roots cause fissures and cracks, which increase the top layer's porosity (Karaca et al., 2018; Sun et al., 2018). As a result, there may be an increase in radon emissions and precipitation percolating into the tailings. There is a tiny chance that radioactivity and trace metals picked up by some plants from the tailings or the cover material will enter the food chain for livestock and people (Carter, 2015).

Although growing greenery and plants over repaired tailings provides numerous benefits, there are also some drawbacks. The top layer becomes more porous as a result of fractures and fissures that vegetation's roots cause (Karaca et al., 2018; Sun et al., 2018). This could boost radon emissions and allow precipitation to seep into the tailings. Some plants take in radioactivity and trace metals from the tailings or the cover material, which has a slim chance of making its way into the food chain for cattle and people (Carter, 2015).

Uptake of radionuclides by these plants and shrubs, including *Cynodondactylon* (doob grass), has been studied (Basu et al., 2000). *Cynodondactylon* shows the highest affinity for Mn and Zn, which are present in the tailings. This is the only component of the vegetation that may be grazed by cattle. All other species mentioned earlier showed a soil to plant transfer factor in the range of 10^{-3} to 10^{-2}. *Typha latifolia* showed relatively higher uptake. Laboratory experiments carried out on vetiver grass (*Vetiveria zyzanioides*, khas grass) have shown high uptake for uranium from uranium tailings (Venu Babu, 2007). After maximum uptake, the vetiver is reported by the researcher to dry up within 35 to 40 days. This appears to have an attractive phytoremediation property. This variety is also known for high soil binding properties. The vegetation or plants that show no significant uptake of radionuclide or trace element may be suitable for phytoremediation over short periods of time, but they may not be suitable for a vegetative cover directly over the tailings due to the probability of radionuclides and trace metals reaching the food chain. However, due to their shallow root penetration and survival under harsh conditions, they are attractive candidates for plantation over remediated and soil-covered tailings. This is true for the vetiver as well (Venu Babu, 2007; Singh and Soni, 2010b).

Studies by Bassegio et al., (2020) suggested that annual sunflowers (Helianthus annus) and Indian mustard (Brassica juncea) might be suitable for phytoremediation, but research from Germany showed that transfer from soil to plant, particularly the roots, was on the order of 103(Huhle et al., 2008).

Thus, field investigations were unable to verify their suitability for efficient phytoremediation. These may be applied to the soil cover over the remedied tailings due to the low or negligible uptake of radionuclides and shallow root penetration (Khan, 2008). Studies at the Indian Institute of Mine, Dhanbad, India, indicated that lemon grass (*Cymbopogon flexuosus*) could be grown on fly ash (Ghosh et al., 1998). As lemon grass also has shallow roots, its suitability for growing over remediated tailings may be studied. To reduce the dispersal of dust into the environment, it is suggested that plantation of Jatropha, a source of biofuel, on the periphery of tailings facilities may be considered as a part of the greenbelt (Khan, 2008). According to

UMH-IV, 2005 operations in eastern Germany have carried out large-scale remediation of waste rocks and uranium mill tailings over a very large area (after successful remediation and vegetative cover, many sites have been released for floriculture, sports and recreational activities).

Research efforts in India need to be accelerated to identify suitable vegetation and plant species that can be used on uranium tailings to provide cover and consolidation of radionuclides and subsequent unrestricted release. Detailed studies on various aspects are suggested, as some plants are known to pick up radionuclides in one system and not in another system. Hence, each tailing site may have varying characteristics with respect to suitability of the particular species of vegetation and plants (Khan, 2008). Transfer factors for radionuclides and trace metals from soil to vegetation and plants are low; therefore, their transfer to animals and humans will be further reduced by a few orders of magnitude. Remediated tailing sites with appropriate cover and growth of suitable vegetation may be considered for release to the public for floriculture, warehousing, tourism and sports activities (Khan, 2008; Soni and Singh, 2011).

7 Case Study
Uranium Tailings in India

In India, uranium mines are operated by the Uranium Corporation of India (UCIL), which is owned by the Atomic Energy Department of India. These are located in the villages of Jaduguda, Bhatin and Narwarpahar in the tribal belt of India known as Jharkhand. The Jaduguda uranium deposit is located almost in the center of the Singhbhum thrust belt. Uranium mineralization in Jaduguda is confined to sheared rock types of Singhbhum thrust zone. Geologically, the thrust belt is constituted by Archean metasediments such as mica-schist, quartzite, phyllites and altered tuffs. The rock types in this zone are broadly classified into two groups—the older Chaibasa stage of rocks consisting of metasediments and the younger Dhanjori stage of rocks consisting of metavolcanics. It has also been observed that there are zones of granular rock and autoclastic conglomerate in the mine where the imprints of shearing are very little. Such zones are normally devoid of uranium. During shearing, the older Chaibasa stages of rocks are thrust over younger metavolcanics of Dhanjori. As a result, the younger Dhanjori stage of rocks lie below the older Chaibasa stage of rocks. The thrust contact between two stages of rocks is severely sheared and brecciaed. Uranium occurs in this sheared zone in very finely disseminated form. The topography of Jaduguda consists of hills and plateaus. All the mountains of Jaduguda are covered with lush green forest. The terrain in general is moderately steep to very steep and extremely rugged (Sarangi, 2003).

All of India's nuclear operations rely on the uranium taken from this area. The ore is mined about 1600 feet (500 meters) below the earth. Nearly 30,000 people in 15 villages live near the mines and the tailing pond that is 100 acres (40 hectares). These tailings, very low in radioactivity but looking into the probable future a hazard threat, need to be restored in an ecologically safe manner so that the plant species used to restore the tailings may both consolidate the radioactivity and pose no health hazard to people living in the area and their livestock. Since these mines are located in an important ecological terrain, the appropriate screening of plant species and their multiplication in uranium tailing areas would help the mining sector, the tribals and the other living entities (both human and animal) in the vicinity as well as the site ecology.

7.1 CASE STUDY AREA AND SELECTION OF SPECIES

The case study on tailing ponds incorporated field-testing of these different methods and costs for the various stabilization procedures. Consolidation of radioactivity in tailings has to be taken up in an ecologically safe manner so that the plants species used not only provide green cover on these tailings but are also both ecologically sustainable and pose no health hazard to people in the area and their livestock. Since these mines are located in an important ecological terrain, the appropriate screening of plant species

DOI: 10.1201/9781003326151-7

in restoration is required to be undertaken so that no uptake of radionuclides by plants occurs. Even if a low amount of uptake occurs, the plants must not be used by people or cattle. This restrains the movement of radionuclides in the food chain. Further, this would help the mining sector, the tribals as well as the site ecology.

The case study presented and discussed aimed at:

- assessing the ecological impact of uranium tailings on soil and vegetation,
- evaluating the radionuclide uptake by vegetation growing on reclaimed tailing sites,
- evaluating the distribution of radionuclides in tailing ponds at different depths, and
- identifying the physicochemical characteristics of soil and tailings.

The study was undertaken to develop a protocol for sustainable stabilization of radioactivity using native species of grass, herbaceous and shrub species. The main objectives the study were:

- selection and propagation of nonedible and nondomestic species with good conservation value,
- phytostabilization of uranium mill tailings to prevent their atmospheric dispersal and adverse impact on the environment, and
- development of greenbelts to ameliorate the tailing sites.

Four sites located at the same altitude and in the same areas were selected for the study: (1) tailing pond; (2) village land; (3) forest land on the periphery of tailing pond, and (4) experimental drum (containers) at the Health Physics Unitcampus in Jaduguda for *ex situ* studies.

Selected plant species were either grown through transplanting seedlings or root stock or sowing seeds over the soil-covered tailing pond for the study of binding capacity, root penetration capacity and radionuclide uptake. An attempt was made to grow the selected species over the soil-covered tailing pond (*in situ* experiment) under laboratory conditions (200 liter containers for *ex situ* experiment), and factors affecting the uptake are also discussed. Evaluation of plant/substrate was an important part of the research. Texture, bulk density, porosity, electrical conductivity, pH, organic carbon, total nitrogen, available phosphorus, exchangeable potassium, natural uranium, polonium and radium were recorded both *in situ* (soil capping on tailing pond) and *ex situ* (experimental containers). The growth of these plants both aboveground and underground (root penetration studies) and radionuclide uptake were researched. The study was undertaken to develop a protocol for sustainable stabilization and consolidation of radioactivity using native species of grass, herbs and shrub species.

The study's primary goals were the selection and multiplication of non-domestic, non-edible species with high conservation value,phytostabilization of uranium mill tailings to stop atmospheric dispersion to reduce negative environmental effects and the creation of greenbelts to improve tailing places.

The climate of the Jaduguda region is rigorous and dry with well-marked seasons representing a typical monsoon-type climate. The winter season extends from December to February; January is the coldest month when the temperature falls 8°C during the night. The summer season extends from March to June. April and May are the hottest months with an average temperature of 33°C. The summer is followed by monsoon that sets by June and lasts until the second week of September or sometimes even up to the end of September. The rainy season is marked by heavy rains, high humidity and a temperature range of 24°C to 32°C. The intensity of rainfall during this period varies between 25 and 60 mm/hour. The average annual rainfall varies from 750 mm to 1300 mm. The forest area is well vegetated. The agricultural land is fertile and crops are irrigated by stream or river waters. The minimum and maximum annual temperature varies from 20°C to 33°C.

The type of soil found in Jaduguda is lateritic, loamy or loamy sand. These are formed from laterite; brown in color chiefly consists of irons and aluminum hydroxides. This type of soil requires an optimum range of temperature for various processes as germination of seeds, growth of roots, absorption of water and solutes, decomposition of organic matter by microorganisms, nitrification, nitrogen fixation, and so on. This type of soil is physiologically dry and generally deep yellow to red. Tailings are of loamy sand/sandy loam texture. Benchmark physical and chemical analysis of soil/tailings prior to experimental trials was done.

Natural forest type in Singhbhum is 5B/CIC (Champion and Seth, 1968) dry peninsular sal under the group of tropical dry deciduous forest. The vegetation of natural sites of village land and forest land include *Shorea robusta*, *Gmelina arborea*, *Syzygium cumini*, *Terminalia tomentosa*, *Terminalia arjuna*, *Madhuca indica*, *Azadirachta indica*, *Mangifera indica*, *Acacia nilotica*, *Adina cordifolia*, *Bauhinia racemosa*, *Bombaxceiba*, *Buchnania lanzen*, *Butea monosperma*, *Cassia fistula*, *Dalbergia sissoo*, *Ficus religiosa*, *Holarrhena antidysenterica* and *Schleichera oleosa*. The undergrowth consists of *Adhatoda zeylanica*, *Zizyphus nummularia*, *Vitex negundo*, *Lantana camara*, *Calotropis procera* and more. A reconnaissance survey of the forests and villages in Jaduguda was done to inventory the plant species of common occurrence.

Although the recorded radioactivity level in these tailings was very low, to avoid any long-term impact on the atmosphere and biosphere, the present study, sponsored by the Board of Research in Nuclear Studies (BRNS) to the Forest Research Institute, was undertaken to find out appropriate plant species of forestry origin for phytostabilizing the tailings and consolidate radioactivity below the soil surface.

A rapid ecological appraisal of the uranium mines and tailing ponds at Jaduguda was taken up. Floristic survey was done and species richness in the adjoining natural forests and tailing ponds was carried out.

Species of ethno botanical relevance have been recorded from the surrounding forest belt. On the basis of floristic surveys, it is concluded that species richness is significantly low on the tailings as compared to natural forest. Therefore, only primary colonizing nonedible and nondomestic species with good conservation value have been selected.

Most plants take up their nutrients and radionuclides mainly via the roots from the soil. The soluble forms of uranium seem to be readily absorbed by the plants.

Radionuclide concentration in plants declines in the order: root > shoot. Revegetation is essentially required to consolidate the radioactivity in the tailings so
that it does not emanate in the atmosphere or enter the food chain. Development
of plant cover has a role in improvement of physico chemical properties of soil
covering the tailings within two years of revegetation. The accumulation and subsequent decomposition of plant residues has resulted inbuilding up of organic carbon with an associated increase in nutrient enrichment. The organic carbon and
nitrogen contents tended to increase with age. The availability of phosphorous and
exchangeable cations showed a marked improvement with the concentration of
potassium increasing progressively. The improvement in these soil factors is very
important for plant growth.

Uptake and distribution of radionuclides in soil capping, plant roots and shoots is
in tailings > soil capping on tailings > roots of selected plant species > shoots of all
the selected species.

On the basis of Phase I of the present study *Jatropha gossypifolia*, *Furcraea
foetida* and *Dodonaea viscosa* are recommended species for consolidating the
radioactivity and thus controlling the adverse impacts on the ecosystem health
as well as the health of human and animals inhabiting the area. Results of Phase
I of the study led to the conclusion that all seven species selected and tried in
the experiments show minimum concentration or below detective limit of the
uptake of radionuclides by plants. This is also because the roots of the selected
species are confined to the top 30 cm layer of soil and do not actually penetrate
the tailings. These results are based on the preliminary investigations of one- and
two-year-old plant trials.

After completion of Phase I, the study was further extended to screen some more
species suitable for vegetation cover on the tailing pond. After extensive literature
survey, 31 species were selected for trial under Phase II of the study. *Ex situ* as well
as *in situ* experiments on the lines of Phase I were conducted for seven months in
Jaduguda as well as at NABTD, Mumbai. Uptake of radionuclides by these species
has also been studied.

Out of 29 species, only 23 species germinated under *ex situ* conditions, while
under *in situ* conditions only 16 species germinated and survived.

Ten species—*Alysicarpus monilifer*, *Argemone mexicana*, *Cenchrus ciliaris*,
Cenchrus setigerus, *Cleome viscosa*, *Crotalaria sericea*, *Croton bonplandianus*,
Desmodium gangeticum, *Dicanthium annulatum* and *Heteropogon contortus*—grew
extremely well under both conditions, whereas *Hyptis suaveolens*, *Indigofera trita*,
Iseilema laxum, *Ocimum canum*, *Rhynchosia minima* and *Tephrosia purpurea* grew
well under *in situ* conditionsonly. However, as the data show, the rooting depth of
three species—*Crotalaria sericea* (23.67 cm.), *Croton bonplandianus* (26.55 cm)
and *Hyptis suaveolens* (24.60cm)—is between 23.67 cm and 26.55cm in a time
period of seven months. The above ground growth of these species is also relatively
high, since the root system may go beyond the soil capping of 30 cm. The species
with deep root systems and higher canopies are not recommended for long-term use
as they may emit radon into the atmosphere (Figure 7.1).

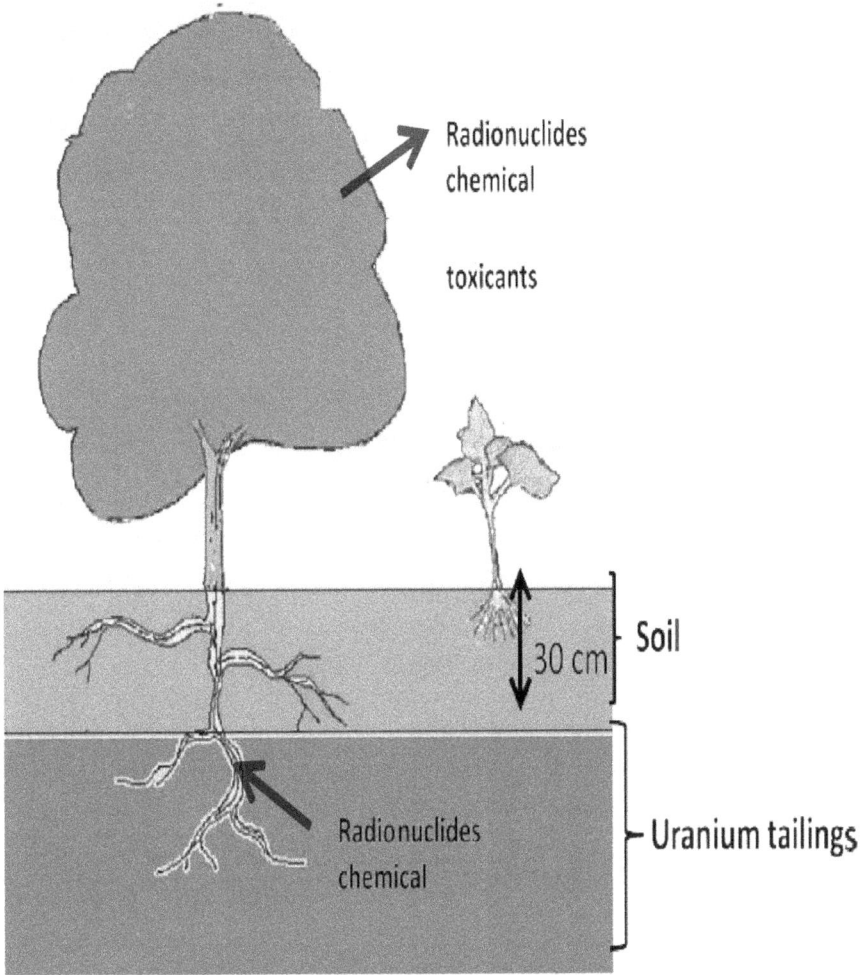

FIGURE 7.1 Plants with a shallow root system and low crown cover are suitable for revegetation of tailings area.

Three experimental sites in Jaduguda in consultation with UCIL and the Health Physics Unit, Government of India were selected (Figure 7.2) for detailed studies and to recommend plant species most suited for consolidation of radioactivity in tailings:

- Tailing pond II (Figure 7.2),
- Periphery of tailing pond II for raising green belt (Figure 7.3),
- *Ex situ* experimental area in the Health Physics Unit, Jaduguda (Figure 7.4), and

FIGURE 7.2   *In situ* experimental trials.

FIGURE 7.3   Forest area on the periphery of tailing pond.

FIGURE 7.4   *Ex situ* experiments.

FIGURE 7.5   Village land in the vicinity of tailing pond.

- Village land in the vicinity of tailing pond II was selected in the one-kilo-meter radius of the tailing pond area. These are flat lands where villagers' cattle graze (Figure 7.5).

7.1.1 *In Situ* Experiments

Tailing pond II at Jaduguda is an engineered impoundment system to practically hold the entire amount of solid waste generated during the processing of uranium ore. After filling the tailing pond with fine slurry, it was capped with 30 cm of soil (Figure 7.2). The tailing pond was capped with a 30 cm thick layer of soil. Seeds of the selected plant species (Tables 7.1 and 7.2) were sown over soil-capped tailings for study of aboveground and underground growth, binding capacity and radionuclide uptake (uranium, polonium and radium) in the first phase and radionuclide uptake (uranium and polonium) in the second phase.

TABLE 7.1
Species of Common Occurrence in the Surrounding Forest Area

Sl.No.	Plant species	Common name	Family
	Trees		
1.	*Acacia nilotica*	Kata	Mimosaceae
2.	*Adina cordifolia*	Karam	Rubiaceae
3.	*Aegle marmelos*	Bel	Rutaceae
4.	*Albizzia lebbeck*	Seris	Mimosaceae
5.	*Artocarpus lakoocha*	Dahu	Artocarpaceae
6.	*Azadirachta indica*	Neem	Meliaceae
7.	*Bauhinia variegata*	Kachnar	Caesalpiniaceae
8.	*Bombax ceiba*	Semal	Bombaceae
9.	*Buchanania lanzen*	Piyal/Chahar	Anacardiaceae
10.	*Dalbergia sissoo*	Shisham	Papilionaceae

(Continued)

TABLE 7.1 (Continued)

Sl.No.	Plant species	Common name	Family
11.	*Ficus cunia*	Podo	Moraceae
12.	*Ficus glabella*	Putkul	Moraceae
13.	*Ficus glomerata*	Gular/loa	Moraceae
14.	*Gmelina arborea*	Gamhar	Verbenaceae
15.	*Holarrhena antidysenterica*	Kurchi	Apocynaceae
16.	*Holoptelia Integrifolia*	Churla	Urticaceae
17.	*Lagerstroemia parviflora*	Secrec	Lythraceae
18.	*Madhuca indica*	Mahua	Sapotaceae
19.	*Mangifera indica*	Aam	Anacardiaceae

The periphery of tailing pond II is only 500 meters away from the tailing pond. The forest is dry peninsular sal under the group of tropical dry deciduous forest. The predominating speciesfoundinthenaturalforestare*Shorearobusta*, *Gmelinaarborea*, *Syzygiumcumini*, *Terminalia arjuna*, *Azadirachta indica*, *Mangifera indica*, *Madhuca indica*, *Schleicheva oleosa* and *Terminalia tomentosa*. The undergrowth consists mainly of *Adhatodazeylanica*, *Zizyphusnummularia*, *Vitexnegundo*, *Lantanacamara* and *Calotropisprocera* (Figure 7.3).

7.1.2 *Ex Situ* EXPERIMENTS

Seeds of the selected plant species (Tables 7.1 and 7.2; Figure 7.4) were sown and in 200-liter experimental containers filled with tailings upto 50 cm and then capped with 30 cm of soil as in the field trials for study of aboveground and underground growth, binding capacity and radionuclide uptake (uranium, polonium and radium) in the first phase and radionuclide uptake (uranium and polonium) in the second phase.

TABLE 7.2

Species of Common Occurrence on Tailing Pond before Soil Capping

Sl.No.	Plant species	Common name	Family
	Shrubs		
1.	*Acacia nilotica*	Kata	Mimosaceae
2.	*Dalbergia sissoo*	Shisham	Papilionaceae
3.	*Pongamia pinnata*	Karanj	Papilionaceae
4.	*Terminalia arjuna*	Arjun	Combretaceae

Sl.No.	Plant species	Common name	Family
	Herbs and Grasses		
1.	*Croton bonplondianum*	Croton	Euphorbiaceae
2.	*Cynodon dactylon*	Dub Grass	Poaceae
3.	*Imperata cylindrica*	Phoola	Poaceae
4.	*Saccharum spontaneum*	Kans	Poaceae
5.	*Typha latifolia*	Typha	Typhaceae

These two types of experimental trials have been conducted to compare growth, binding capacity, root penetration and uranium uptake by plant species under both *in situ* and *ex situ* conditions.

Village land in the one-kilometer radius of the tailing pond area was selected for ethno botanical studies. These are flat lands where villager's cattle graze. On the basis of ethno botanical survey in the villages and in adjoining forests, nonedible and non-domestic species with good conservation value were selected for experimental trials.

Floristic survey of adjoining natural forests and tailing pond prior to spread of soil cover on the tailing pond was carried out tore cord the species richness in these areas. For this, a rapid ecological appraisal of the two sites was taken up. The forest type surrounding tailing pond is peninsular dry sal 5B/ic tropical dry deciduous forest (Champion and Seth, 1968). Total number of species of common occurrence recorded is 35, which belong to 25 families. There were 24 tree, six shrub and five herb aceous species.

Floristic survey of the tailing pond carried out in between to assess the biological spectrum of species occurring on the tailing pond showed that no trees of natural occurrence were found on the tailings; however, four species of planted trees and five herb and grass species could be recorded (Table 7.2).

The floristic surveys led to the conclusion that species richness is significantly low on the tailings as compared to natural forest, which means that these tailings do not provide a suitable strata for vegetation growth. Twelve indigenous species (shrubs and grasses) from nearby sites that do not have deep roots and high crown but have good conservation value were initially selected and planted on the tailing ponds. These species have been selected after undertaking the ethnobotanical survey of five tribal village areas surrounding the tailing pond that clearly showed that these species do not have any value for the villagers.

7.1.2.1 Forest Species of Ethnobotanical Relevance

Jharkhand state is one of the biodiversity-rich regions of the country wherein large number of plant species have been in use by ethnic communities. There are 30 tribes and subtribes in the Jharkhand region. The major tribes in and around Jaduguda are Santhals, Mundas and Hos. A total of 81 species that are being collected by local villagers for medicinal and other uses are listed in Table 7.3 and have been recorded on the basis of ethnobotanical survey undertaken in the surrounding villages. These surveys were undertaken to form the basis of selection for primary colonizing native plant species of forestry origin.

The main criteria for species selection have been their growth form, low crown cover, shallow root system, nonedible nature, non domestic use and good conservation value (Table 7.4).

TABLE 7.3
List of Plant Species of Ethnobotanical Importance

S. No.	Botanical name	Local name	Family	Part used	Habit
1.	*Abrus precatorius*	Rati	Fabaceae	Seeds, leaves and roots	Climber
2.	*Acacia concinna*	Shikakaaee	Mimosaceae	Leaves, bark and pod	Shrubs
3.	*Acacia nilotica ssp. indica*	Babul	Mimosaceae	All parts	Tree
4.	*Achyranthus aspera*	Latjira	Amaranthaceae	All parts	Herb
5.	*Adhatoda vasica*	Vasaka	Acanthaceae	Root, leaves and flower	Large shrub
6.	*Adiantum ilunulatum*	Maiden hair fern	Adianataceae	All plants	Herb
7.	*Andrograhis paniculata*	Kalmegh	Acanthaceae	Leaves and fruits	Herb
8.	*Annona squamosa*	Ata	Annonaceae	Leaves	Tree
9.	*Aegle marmelos*	Bel	Rutaceae	Fruit, root bark, leaves	Tree
10.	*Ageratum conyzoides*	Ghaneraosadi	Asteraceae	Leaves	Herb
11.	*Alangium salvifolium*	Ankul	Alangiaceae	Bark and fruits	Shrub
12.	*Aloe barbadensis*	Kumaree	Liliaceae	Pulp, dried leaves	Shrub
13.	*Alstonia scholaris*	Chatani	Apocynaccac	Bark	Tree
14.	*Andrographis paniculata*	Kalmegh	Acanthaceae	All parts	Herb
15.	*Anogeissus latifolia*	Dhaurea gach	Combretaceae	Bark and wood	Tree
16.	*Asparagus racemosus*	Satavari	Asparagaceae	Root and leaves	Herbs
17.	*Azadirachta indica*	Neem	Meliaceae	All parts	Tree
18.	*Bacopa monnieri*	Bramhi	Scrophularriaceae	Stem and leaves	Herb
19.	*Basella alba*	Potaki	Basellaceae	Root, stem and leaves	Climber

S. No.	Botanical name	Local name	Family	Part used	Habit
20.	Bauhinia malabarica		Caesalpiniaceae	Bark	Tree
21.	B. vahlii	Room	Caesalpiniaceae	Leaves	Climber
22.	B. variegata	Kachnara	Caesalpiniaceae	Bark, root and gum	Tree
23.	Boerhavia diffusa	Punarnawa	Nyctanthaceae	Whole plant	Herbs
24.	Buchanania lanzan	Chironji	Anacardiaceae	Fruit, bark, leaves and oil	Tree
25.	Butea monosperma	Palas	Fabaceae	Gum, seed, flower, leaves and bark	Tree
26.	Calotropis giganteus	Madar	Asclepiadaceae	Roots, flower and leaves	Small tree
27.	Calotropis procera	Aak	Asclepiadaceae	Leaves	Shrub
28.	Carica papaya	Papaya, Papaw	Caricaceae	Latex	Shrub
29.	Cassia alata	Dadmari	Caesalpiniaceae	Leaves andflower	Shrubs
30.	Cassia fistula	Sonari gach	Caesalpiniaceae		Tree
31.	Centella asiatica	Brahma manduk	Apiaceae	Leaves	Herbs
32.	Cissus quadrangularis	Haadajoda	Vitaceae	Root, stem and leaves	Climber
33.	Cleome viscose	Wild mustard	Cleomaceae	Seed and leaves	Herbs
34.	Clerodendrum serratum	Barangee	Verbenaceae	Root, leaves and seeds	Shrubs
35.	Croton tiglium	Jamalgota	Euphorbiaceae	Seed and oil	Shrubs
36.	Curculigo orchioides	Kaali musali	Hypoxidaceae	Tuberous roots and pulp	Herbs
37.	Curcuma amada	Amhaldi	Zingiberaceae	Root and tuber	Herbs
38.	Curcuma aromatica	Junglihalii	Zingiberaceae	Tuber and rhizome	Herb
39.	Curcuma longa	Haldi	Zingiberaceae	Tuber and rhizome	Shrub
40.	Cuscuta reflexa	Nirmuli	Cuscutaceae	Tuber and rhizome	
41.	Desmodium cephaloides	Ram datum	Fabaceae	Leaves, bark and stem	Herb
42.	Dillenia aurea	Rye	Dilleniaceae	Fruits pickled	Tree
43.	Dolichos biflorus	Jugle kulthi	Fabaceae	Seeds	Herb
44.	Echinops echinatus	Gokru	Asteraceae	Fruits and whole plants	Herb
45.	Euphorbia-hirta	Dudhi	Euphorbiaceae		Herb

(Continued)

TABLE 7.3 (Continued)

S. No.	Botanical name	Local name	Family	Part used	Habit
46.	Ficus racemosa	Gular	Moraceae	Root, bark, fruit and latex	Tree
47.	Helicteres isora	Murmuri	Sterculiaceae	Fruit juice	Shrub
48.	Holarrhena antidysentrica	Kurchi	Apocyanaceae	Bark, seeds and leaves	Tree
49.	Hygrophilla spinosa	Gokshura	Acanthaceae	Root, seed and leaves	Herbs
50.	Ichnocarpus frutesens	Kallidudhi	Apocynaceae	Roots and leaves	Shrub
51.	Indigofera cassioides	Gilori	Fabaceae	Flowers	Shrub
52.	Jatropha curcas	Longta halo	Euphorbiaceae	Latex, fruits	Small tree
53.	Lawsonia inermis	Mehandi	Lythraceae	Leaf, and bark flower	Shrubs or small tree
54.	Madhuca indica	Mahua	Sapotaceae	Flower, bark, leaves and seeds	Tree
55.	Mimosa pudica	Lajwanti	Mimosaceae	Leaves, root and seed	Herb
56.	Moringa oleifera	Munga ara	Moringaceae	Root paste fruits leaves, flowers	Tree
57.	Nyctanthus arbor-tristis	Harsingar	Oleaceae	Whole plants	Small tree
58.	Ocimum graitssimum	Bantulsi	Lamiaceae	Leaves and seeds	Shrubs (A)
59.	Phyllanthus emblica	Amla	Euphorbiaceae	Dried fruit, seed, leaves and bark	Tree (P)
60.	Phyllantus niruri	Jar-Ambla	Euphorbiaceae	Whole herbs	Herb (A)
61.	Plumbago zeylanica	Chitrak	Plumbaginaceae	Leaves, root and bark	Herb (A)
62.	Rauwolfia serpentina	Patal garnd	Apocynaceae		Herb
63.	Ricinus communis	Arandi	Euphorbiaceae	Chota Patal Garnd	shrub
64.	Saraca asoka	Sita-ashok	Caesalpiniaceae	Leaf, seed androots	Tree (P)
65.	Schleichera oleosa	Kusum	Oleaceae		Tree
66.	Semecarpus anacardium	Bhela	Anacardiaceae	Fruit, oil and godambi	Tree (P)
67.	Shorea robusta	Sal	Dipterocarpaceae	—	Tree
68.	Sida acuta	Naagbala	Malvaceae	Root, seed and leaves	Herbs (P)

S. No.	Botanical name	Local name	Family	Part used	Habit
69.	Sida cordifolia	Bala	Malvaceae	Root, seed and leaves	Herb
71.	Terminalia arjuna	Arjuna	Combretaceae	Bark	Tree (P)
72.	Terminalis belerica	Baheraa	Combretaceae	Bark	Tree (P)
73.	Terminalis chebula	Harra	Combretaceae	Fruits	Tree (P)
74.	Terminalia tomentosa	Asan	Combretaceae	Bark	Tree (P)
75.	Thysanolaena agrostis	Thugach	Poaceae	Stem inflorescence	Grass
76.	Tridex	Tridex	Asteraceae	Leaves	Herb (P)
77.	Trifolium sp.	Jungle Kurt	Fabaceae	Root paste applied above the ear onthe back	Herb
78.	Triumfetta rhomboidea	Chikti	Tiliaceae	Fruit, flower and bark, leaves	Herbs (A)
79.	Vitex negundo	Nirgundi	Verbenaceae	Leaves	Large shrubs (P)
80.	Withnania coagulans	Aswagandha	Solanaceae	Leaves, roots	Shrubs (A)
81.	Xanthium strumarium	Stem and leaves	Asteraceae	Medicinal	Shrub

TABLE 7.4
List of Primary Colonizing Species Screened for Trial in Phase I on the Basis of Their Non-Ethnobotanical Value

Sl.No.	Plant species	Family	Habit	Common name
1.	*Arundo donax	Poaceae	Grass	Nal
2.	Colebrookeaoppositifolia	Lamiaceae	Shrub	Binda
3.	Dodonaea viscosa	Sapindaceae	Shrub	Jangali mehandi
4.	Imperata cylindrica	Poaceae	Grass	Phoola
5.	*Ipomoea fistulosa	Convolvulaceae	Shrub	Beha
6.	Furcaria foetida	Agavaceae	Shrub	Furcaria
7.	Jatropha gossypifolia	Euphorbiaceae	Shrub	Red Jatropha, Pahar
8.	*Jatropha curcas	Euphorbiaceae	Shrub	Jatropha
9.	*Phragmites karka	Poaceae	Grass	Nali
10.	Pogostemon benghalense	Lamiaceae	Herb	Phangla,Jullata
11.	*Rumex hastatus	Polygonaceae	Herb	Chulmora, khatta
12.	Saccharumspontaneum	Poaceae	Grass	Kans

Note:* Species that did not survive/deep root system on tailing pond area.

TABLE 7.5

Selected Plant Species for Experimental Trials in Phase I after Preliminary Screening

Sl.No.	Plant species	Family	Common name	Plant form
1.	*Colebrookea oppositifolia*	Lamiaceae	Binda	Shrub
2.	*Dodonaea viscosa*	Sapindaceae	Vilayati Mehandi	Shrub
3.	*Furcraea foetida*	Agavaceae	Furcaria	Shrub
4.	*Imperata cylindrica*	Poaceae	Phoola	Grass
5.	*Jatropha gossypifolia*	Euphorbiaceae	Red Jatropha	Shrub
6.	*Pogostemon benghalense*	Lamiaceae	Phangla,Julata	Shrub
7.	*Saccharum spontaneum*	Poaceae	Kans	Grass

- After preliminary trials, five species out of 12 did not survive and hence were not tried further. Other criteria for selection are their growth form *viz.*, perennials with low crown cover, shallow root system and conservation value of the species. Seven species were selected and tried for *in situ* as well as *ex situ* experiments (Table 7.5) during Phase I of the project. The other important criteria were:
- species that can provide a sustainable and productive vegetative cover,
- species that do not have any socioeconomic relevance in the area,
- species with low radionuclide uptake to avoid its dissemination in any form in the environment/food chain, and
- species with high conservation value.

After successful trial of seven species under *in situ* and *ex situ* experiments during Phase I, further experiments were extended on 31 species. During Phase II of the project, a total of 32 species were selected for *in situ* as well as *ex situ* experiments (Table 7.6).

TABLE 7.6

Selected Plant Species for Experimental Trials in Phase II

S.No.	Botanical name	Family	Common name	Plant form
1.	*Alysicarpus monilifer*	Fabaceae	Leel	Herb
2.	*Argemone mexicana*	Papaveraceae	Satyanashi	Herb
3.	*Bothriochloa intermedia*	Poaceae	Pita grass	Grass
4.	*Bothriochloa pertusa*	Poaceae	Pita grass	Grass
5.	*Celocia urgentea*	Amaranthaceae	Chilbill	Herb
6.	*Cenchrus ciliaris*	Poaceae	Anjan grass	Grass
7.	*Cenchrus setigerus*	Poaceae	Dhamin grass	Grass
8.	*Chrysopogon fulvus*	Poaceae	Guria grass	Grass
9.	*Cleome viscosa*	Cleomaceae	Harhara	Herb

S.No.	Botanical name	Family	Common name	Plant form
10.	*Clitoria ternatea*	Fabaceae	Aparjita	Herb
11.	*Crotalaria juncea*	Fabaceae	Sunn	Shrub
12.	*Crotalaria sericea*	Fabaceae	Junka	Shrub
13.	*Croton bonplandianus*	Euphorbiaceae	Jamal ghota	Herb
14.	*Cymbopogon citratus*	Poaceae	Lemon grass	Grass
15.	*Desmanthus virgatus*	Fabaceae	Dasrath ghas	Shrub
16.	*Desmodium gangeticum*	Fabaceae	Shalaparni	Herb
17.	*Desmostachya bipinnata*	Poaceae	Durva	Grass
18	*Dicanthium annulatum*	Poaceae	Marvel grass	Poaceae
19.	*Heteropogon contortus*	Poaceae	Pili grass	Poaceae
20.	*Hyptis suaveolens*	Lamiaceae	Ganga tulsi	Shrub
21.	*Indigofera trita*	Fabaceae	Indigo	Herb
22.	*Iseilema laxum*	Poaceae	Machauri grass	Grass
23.	*Mucuna pruriens*	Fabaceae	Velvet bean	Herb (creeper)
24.	*Ocimum canum*	Lamiaceae	Jangali tulsi	Herb
25.	*Panicum antidotale*	Poaceae	Panic grass	Grass
26.	*Pennesetum caudatum*	Poaceae	Fountain grass	Grass
27.	*Rhynchosia minima*	Fabaceae	Nilphali	Herb (creeper)
28.	*Ruellia tuberosa*	Acanthaceae	Popping pod	Herb
29.	*Stylosanthes hamata*	Fabaceae	Stylo hamata	Herb
30.	*Tephrosia purpurea*	Fabaceae	Ban nil	Herb
31.	*Thysanolaena maxima*	Poaceae	Jharu grass	Grass
32.	*Vetiveria zyzanioides*	Poaceae	Khas grass	Grass

7.2 ABOVEGROUND AND UNDERGROUND PLANT GROWTH PERFORMANCE

Seed material of 32 species (Table 7.6) was sown in thermocol cups containing soil-rite artificial soil medium. Each species was sown in excess, in six to eight cups, as the seed material was collected from the wild and the information on germination percentage and dormancy was not known. The cups were watered and left in growth racks with a 12-hour photoperiod and a temperature of 25–27°C. After germination, wherever required, thinning was done, leaving four to six plants in each cup. The growth of the seedlings was supported by Steinberg's liquid nutrient medium once in three to four days or as often as required when the seedlings grew bigger.

Out of 31 species selected under Phase II of the study, only 25 species germinated under *in situ* conditions while only 18 species germinated under *ex situ* conditions. However, under natural conditions all the 31 species were recorded growing well. Ten—*Alysicarpus monilifer, Argemone mexicana, Cenchrus ciliaris, Cenchrus setigerus, Cleome viscosa, Crotalaria sericea, Croton bonplandianus, Desmodium gangeticum, Dicanthium annulatum* and *Heteropogon contortus*—grew extremely well under both conditions, whereas *Hyptis suaveolens, Indigofera trita, Iseilema laxum, Ocimum canum, Rhynchosia minima* and *Tephrosia purpurea* grew well under *in situ* conditions only. However, as the data show, the rooting depth of three species—*Crotalaria sericea* (23.67 cm), *Croton bonplandianus* (26.55 cm) and *Hyptis suaveolens* (24.60 cm)—is between 23.67 cm and 26.55 cm in a time period of seven months. The aboveground growth of these species is also relatively high, but since the root system may go beyond the soil capping of 30 cm, the species are not recommended for long-term use.

Mucuna pruriens, which is a climber, showed maximum height (length) of 396 cm followed by *Pennesetum caudatum* (106.8), *Bothriochloa intermedia* (110.80), *Panicum antidotale* (120.4) and *Chrysopogon fulvus* (146.00). Lowest height recorded was for *Alysicarpus monilifer* (33.20).

Hyptis suaveolens, Indigofera trita, Iseilema laxum, Ocimum canum, Rhynchosia minima and *Tephrosia purpurea* did not grow under *ex situ* conditions. *Stylosanthes hamata, Clitoria ternatea, Vetiveria zizanioides, Thysanolaena maxima, Desmanthus virgatus, Bothriochloa pertusa, Cymbopogon citratus, Crotalaria juncea, Pennesetum caudatum, Bothriochloa intermedia, Panicum antidotale, Chrysopogon fulvus* and *Mucuna pruriens* did not grow under *in situ* conditions.

Maximum collar diameter was recorded for *Croton bonplandianus* and *Argemone Mexicana*, while the minimum recorded was for *Bothriochloa intermedia* and *Pennesetum caudatum* under *ex situ* conditions. While under *in situ* conditions, *Croton bonplandianus* and *Argemone Mexicana* had maximum recorded diameter, and the minimum was for *Heteropogon contortus* and *Cenchrus ciliaris*.

TABLE 7.7
Recommended Plant Species for Tailing Pond

S. No.	Plant species	Family
1.	*Alysicarpus monilifer*	*Fabaceae*
2.	*Argemone mexicana*	*Papaveraceae*
3.	*Cenchrus ciliaris*	*Poaceae*
4.	*Cenchrus setigerus*	*Poaceae*
5.	*Cleome viscosa**	*Capparidaceae*
6.	*Desmodium gangeticum*	*Fabaceae*
7.	*Dicanthium annulatum*	*Poaceae*
8.	*Heteropogon contortus*	*Poaceae*
9.	*Indigofera trita*	*Fabaceae*
10.	*Iseilema laxum*	*Poaceae*
11.	*Ocimum canum*	*Lamiaceae*
12.	*Rhynchosia minima*	*Fabaceae*
13.	*Tephrosia purpurea*	*Fabaceae*

Note: *Seeds used for culinary purpose.

However, all 29 species were growing well in terms of height and diameter under natural conditions. Comparatively low diameter and height under experimental conditions may be due to the fact that plant growth was recorded for only seven-month-old plants.

To conclude, only the following 13 species have been screened on the basis of growth study. Further, the ethnobotanical study in and around Jaduguda shows that except for *Cleome viscosa*, locals do not use any of the species for their day-to-day requirements. To avoid translocation of radionuclides through the food chain, trial of *Cleome viscosa* is also not recommended since seeds of *Cleome viscosa* are used for culinary purposes. Therefore, out of 13 species listed in Table 7.7, only 12 are safely recommended for revegetating the tailing pond.

7.2.1 HARDENING AND TRANSPLANTATION

Since the seedlings were grown under laboratory conditions, with nutrient medium, hardening was carried out before final transplantation into mill tailings. The thermocol cups containing three-month-old plants were transferred to a net house covered with polythene sheet to prevent rainwater. The plants were continued to be given nutrient solution almost once in two to three days during the process of hardening and until transplanted into mill tailings, that is, for a period of 40 days. Mill tailings fromtailing pond II were obtained from UCIL, Jaduguda. The top layer containing soil covering was removed and the mill tailings from the layers underneath was collected, transported to Bhabha Atomic ResearchCenter (BARC), Mumbai, and stored in a secure place. Just before use, the mill tailings contained in the plastic tins were mixed thoroughly using a wooden stick and several portions were removed from the container and layered over the polythene sheet. Once again, the mill tailings were mixed evenly on the polythene sheet and transferred to 500 ml glass beakers wrapped on the sides and bottom with two layers of carbon paper to avoid algal growth during the experimental period. All the operations involving mill tailings were carried out using gloves.

The plants now aged four months and ten days were dislodged from the thermocol cups gently and the adhering soilrite medium was washed off under a steady stream of tap water, taking care not to separate the entangled roots from different plants, thus avoiding damage to the root system. The bunch of plants (three to five) from each thermocol cup were transferred intact to a single beaker that was partly filled with mill tailings beforehand. The plants were placed in the middle of beakers and filled with further addition of mill tailings, leaving clearance of approximately 1.5 cm from the top of the beakers. The plants were irrigated only with tap water and no additional nutrient solution was supplemented during the remaining period. For every species in mill tailings, comparable controls were maintained in plastic pots containing normal soil mixed with farmyard manure. Watering of plants in both mill tailings and normal soil was carried out approximately once in two to three days depending on the prevailing weather conditions. Every time, just enough water was added to avoid stagnation in the mill tailings as there were no outlets in the beakers, a deliberate procedure followed to prevent leaching of chemical and radiological constituents.

Several species of grasses and legumes could indeed complete their life cycle, including flowering and seed setting. Further, the roots of several legumes could establish well and produce nodules. Based on the vigor of growth, stature of plant and so on, the results of the screening experiment for survival were tabulated (see later). Of the 20 species germinated, nine appeared to be well suited while seven more species shown potential to be used in combination with other species for growth on mill tailings. The remaining four species were found to be unfit for the purpose. Mill tailings need to be managed in such a way that the entire area becomes self-sustaining over a period of time. There have been several advances made in the field of phytoremediation, and the judicious use of the technology can bring dividends in the long run. When it comes to mill tailings, perhaps phytostabilization may be of good application (Mendez and Maier, 2008).

Phytostabilization involves adsorption and/or sequestering of metals and trace elements on the root surface or in the rhizosphere. Further, the elements may also accumulate in the roots with insignificant quantities mobilized to shoots. Thus, the phytostabilization technology not only ensures prevention of soil erosion from wind and rainwater, thereby protecting the soil cover of mill tailings, it also ensures less exposure of livestock, wildlife and humans to the mill tailings.

1. The species of plants chosen should have the ability to withstand the physicochemical and the radiological conditions of the mill tailings.
2. The concentration of the metal/trace element in the aboveground vegetation should not exceed the domestic animal toxicity limits (NRC, 2005a).
3. The metal/radionuclide concentration should not exceed the leaf/tissue toxicity limits.
4. Appropriate seeding or transplantation needs to be carried out for better establishment of the plant community.
5. Periodic irrigation and addition of some amount of manure or limited fertilization need to be applied until the plants are established (Munshower, 1994; Tordoff et al., 2000; Williams and Currey, 2002).
6. Some amount of cultural/agronomic practices, using either herbicides or physical removal of unwanted, especially tree, species need to be taken up with some defined periodicity. Otherwise, such unmanaged mill tailings can get colonized by dominant plant species in the long run.
7. Plant species that have higher leaf area index and can withstand field fires and drought need to be incorporated in the plant community.

7.2.2 HARVESTING OF ROOT AND SHOOT SAMPLES

At the end of the experimental period, the species exhibited significant differences in their growth vigor in mill tailings, and the morphological condition of the plants at the time of harvest is described in Tables 7.8 to 7.11. The total height or length of plants was recorded. The twiners were stretched and the maximum length from the base of the stem to the base of the terminal bud, be it on the main stem or the longest branch, was recorded. In the case of grasses/monocots, the length was measured from the base of the plant to the tip of the spike. Number of tillers and number of branches were recorded wherever applicable. However, in a few grass species, which gave new tillers or bunches of growth from auxiliary buds along the length of the main stem, such appendages were also counted as tillers.

TABLE 7.8
Comparative Growth Performance of Selected Species in *in Situ* Experiments

S.No.	Species	Height (cm)		Diameter (cm)		Basal area cm^2	
		I Yr	II Yr	I Yr	II Yr	I Yr	II Yr
1.	*Colebrookea oppositifolia*	72	120	1.20	1.40	1.13	1.54
2.	*Dodonaea viscosa*	76	90	1.10	1.46	0.95	1.67
3.	*Furcraea foetida*	38	66	1.14	1.42	1.02	1.58
4.	*Imperata cylindrica*	20	30	0.22	0.28	0.04	0.06
5.	*Jatropha gossypifolia*	80	110	1.18	1.46	1.09	1.67
6.	*Pogostemon benghalense*	90	140	1.12	1.40	0.98	1.54
7.	*Saccharum spontaneum*	28	46	0.12	0.22	0.01	0.04

TABLE 7.9
Comparative Growth Performance of Selected Species in *Ex situ* Experiments Height (cm)

S.No.	Species	Height (cm)		Diameter (cm)		Basal area cm^2	
		I Yr	II Yr	I Yr	II Yr	I Yr	II Yr
1.	*Colebrookea oppositifolia*	116.00	140	1.6	2	2.01	3.14
2.	*Dodonaea viscosa*	86.00	110	1.16	1.3	1.06	1.33
3.	*Furcraea foetida*	50.00	70	1.26	1.3	1.25	1.33
4.	*Imperata cylindrica*	26.00	30	0.24	0.34	0.05	0.09
5.	*Jatropha gossypifolia*	90.00	160	1.4	1.64	1.54	2.11
6.	*Pogostemon benghalense*	114.00	166	1.16	1.4	1.06	1.54
7.	*Saccharum spontaneum*	36.00	56	0.2	0.26	0.03	0.05

TABLE 17.10
Average Plant Height (cm) on Different Experimental Sites in Phase II

S.No.	Plant species	*Ex situ*	*In situ*	Natural condition
1.	*Alysicarpus monilifer*	**33.20**	**27.20**	42.24
2.	*Argemone mexicana*	**42.20**	**44.39**	56.40
3.	*Bothriochloa intermedia*	110.80	0.00	127.20
4.	*Bothriochloa pertusa*	76.00	0.00	87.72
5.	*Cenchrus ciliaris*	**88.40**	**34.05**	88.88
6.	*Cenchrus setigerus*	**87.20**	**74.80**	87.40
7.	*Chrysopogon fulvus*	146.00	0.00	144.80
8.	*Cleome viscosa*	**64.00**	**55.00**	56.00

(Continued)

TABLE 7.10   (Continued)

S.No.	Plant species	Ex situ	In situ	Natural condition
9.	Clitoria ternatea	39.00	0.00	54.40
10.	**Crotalaria sericea**	**86**	**170**	200.8
11.	Crotalaria juncea	91.2	0.00	106
12.	**Croton bonplandianus**	**65.248**	**64**	69.4
13.	Cymbopogon citratus	89.4	0.00	156
14.	Desmanthus virgatus	66.8	0	135.6
15.	**Desmodium gangeticum**	**58.4**	**32.8**	74.8
16.	**Dicanthium annulatum**	**42.8**	**22.4**	57.2
17.	**Heteropogon contortus**	**42**	**34.8**	45.6
18.	Hyptis suaveolens	0	84.4	81
19.	Indigofera trita	0	68.8	77.6
20.	Iseilema laxum	0	42	35.6
21.	Mucuna pruriens	396	0	409.6
22.	Ocimum canum	0	40.58	40.6
23.	Panicum antidotale	120.4	0	213
24.	Pennesetum caudatum	106.8	0	107.6
25.	Rhynchosia minima	0	38	77.2
26.	Stylosanthes hamata	33.6	0	35.6
27.	Tephrosia purpurea	0	48	53.6
28.	Thysanolaena maxima	42.8	0	78
29.	Vetiveria zizanioides	39	0	157.6

Note: Species in bold germinated and survived under both *ex situ* as well as *in situ* conditions.

TABLE 7.11
Average Shoot Diameter (mm) on Different Experimental Sites in Phase II

S.No.	Plant species	Ex situ	In situ	Natural condition
1.	**Alysicarpus monilifer**	**1.12**	**1.05**	**1.28**
2.	**Argemone mexicana**	**3.88**	**4.03**	**4.28**
3.	Bothriochloa intermedia	0.84	0	0.92
4.	Bothriochloa pertusa	0.96	0	1.04
5.	**Cenchrus ciliaris**	**0.87**	**0.84**	**0.94**
6.	**Cenchrus setigerus**	**0.9**	**0.87**	**1.02**
7.	Chrysopogon fulvus	0.9	0	1.04
8.	**Cleome viscosa**	**1.99**	**1.23**	**2.66**
9.	Clitoria ternatea	1.64	0	2.04

S.No.	Plant species	*Ex situ*	*In situ*	Natural condition
10.	*Crotalaria juncea*	2.24	0	2.86
11.	**Crotalaria sericea**	**1.64**	**2.57**	**3.08**
12.	**Croton bonplandianus**	**3.9**	**3.87**	**4.04**
13.	*Cymbopogon citratus*	1.18	0	1.44
14.	*Desmanthus virgatus*	1.38	0	1.4
15.	**Desmodium gangeticum**	**1.44**	**1.2**	**1.32**
16.	**Dicanthium annulatum**	**0.98**	**0.93**	**1.2**
17.	**Heteropogon contortus**	**0.94**	**0.82**	**1.08**
18.	*Hyptis suaveolens*	0	2.62	3.86
19.	*Indigofera trita*	0	1.28	1.35
20.	*Iseilema laxum*	0	0.96	1.04
21.	*Mucuna pruriens*	2.18	0	3.82
22.	*Ocimum canum*	0	2.12	2.56
23.	*Panicum antidotale*	2.14	0	2.18
24.	*Pennesetum caudatum*	0.86	0	1.04
25.	*Rhynchosia minima*	0	1.15	2.18
26.	*Stylosanthes hamata*	1.64	0	2.12
27.	*Tephrosia purpurea*	0	2.92	3.28
28.	*Thysanolaena maxima*	0.92	0	1.12
29.	*Vetiveria zyzanioides*	0.89	0	1.1

Note: Species in bold germinated and survived under both *ex situ* as well as *in situ* conditions.

7.2.3 ROOT PENETRATION AND SOIL BINDING CAPACITY OF THE PLANT SPECIES OF EXPERIMENTAL TAILING POND AND CONTAINERS

Plant species under trial in containers and plants grown on a tailing pond were used for studying the periodic growth of these plants both above ground and underground (root penetration) and radionuclide uptake to compare the growth performance with respect to total number of lateral roots, length and diameter of tap root, average root diameter, total root weight and volume of total root system. Root penetration of *Pogostemon benghalense* and *Jatropha gossypifolia* is highest in natural as well as tailing ponds as well as in containers as compared to other species *viz.*, *Dodonaea viscosa*, *Colebrookea oppositifolia*, *Saccharum spontaneum* and *Imperata cylindrica*. Volume of pit left after excavation of complete root system was also assessed. Roots were excavated by mechanical method (Bohm, 1979). Root diameter and root length was measured (Tables 7.12 and 7.13) with the help of measuring tape and vernier caliper respectively, and root volume by water displacement method. Binding capacity of roots of tried species after one and two years has been estimated (Tables 7.14 and 7.15).

TABLE 7.12
Root Diameter of Selected Species at Experimental Sites

S. No.	Root diameter (mm)			
	Plant species	Ex situ	In situ	Natural forest
1.	*Alysicarpus monilifer*	1.08	1.052	0.976
2.	*Argemone mexicana*	5.614	4.032	4.148
3.	*Bothriochloa intermedia*	0.8	0	0.81
4.	*Bothriochloa pertusa*	0.912	0	0.902
5.	*Cenchrus ciliaris*	0.748	0.842	0.890
6.	*Cenchrus setigerus*	1.082	0.872	0.820
7.	*Chrysopogon fulvus*	0.892	0	0.838
8.	*Cleome viscosa*	1.380	1.23	1.224
9.	*Clitoria ternatea*	1.632	0	1.228
10.	*Crotalaria sericea*	2.064	2.572	2.672
11.	*Crotalaria juncea*	1.780	0	2.092
12.	*Croton bonplandianus*	3.524	3.876	3.792
13.	*Cymbopogon citratus*	1.140	0	1.348
14.	*Desmanthus virgatus*	1.328	0	1.388
15.	*Desmodium gangeticum*	1.376	1.208	1.300
16.	*Dicanthium annulatum*	1.186	0.932	1.196
17.	*Heteropogon contortus*	0.844	0.816	0.898
18.	*Hyptis suaveolens*	0	2.436	2.472
19.	*Ingigofera trita*	0	0.530	1.126
20.	*Iseilma laxum*	0	0.952	0.964
21.	*Mucuna pruriens*	1.220	0	1.204
22.	*Ocimum canum*	0	1.156	1.218
23.	*Panicum antidotale*	1.532	0	2.044
24.	*Pennesetum caudatum*	0.832	0	0.926
25.	*Rhynchosia minima*	0	0.960	1.098
26.	*Stylosanthes hamata*	1.068	0	0.546
27.	*Tephrosia purpurea*	0	1.820	1.880
28.	*Thysanolaena maxima*	0.86	0	0.96
29.	*Vetiveria zyzanioides*	0.850	0	0.908

TABLE 7.13
Root Length of Selected Species at Experimental Sites in Phase II

S. No.	Plant species	Main root length (cm)		
		Ex situ	In situ	Natural forest
1.	Alysicarpus monilifer	16.60	13.95	15.79
2.	Argemone mexicana	18.468	13.372	16.74
3.	Bothriochloa intermedia	12.0		13.12
4.	Bothriochloa pertusa	10.44		12.38
5.	Cenchrus ciliaris	10.16	8.34	11.96
6.	Cenchrus setigerus	12.00	8.62	13.16
7.	Chrysopogon fulvus	14.48		14.08
8.	Cleome viscosa	22.4	18.18	24.056
9.	Clitoria ternatea	14.84		16.8
10.	Crotalaria sericea	21.68	23.67	24.56
11.	Crotalaria juncea	21.172		21.906
12.	Croton bonplandianus	24.44	26.55	29.6
13.	Cymbopogon citratus	14.8		18.26
14.	Desmanthus virgatus	19.11		22.14
15.	Desmodium gangeticum	23.48	15.55	23.538
16.	Dicanthium annulatum	12.00	8.056	14.6
17.	Heteropogon contortus	14.4	12.88	16.47
18.	Hyptis suaveolens		24.64	25.164
19.	Ingigofera trita		18.168	19.404
20.	Iseilma laxum		14.266	13.26
21.	Mucuna pruriens	21.432		22.014
22.	Ocimum canum		15.948	16.318
23.	Panicum antidotale	12.4		18.24
24.	Pennesetum caudatum	12.28		13.56
25.	Rhynchosia minima		14.8	15.24
26.	Stylosanthes hamata	14.4		14.346
27.	Tephrosia purpurea		26.64	26.83
28.	Thysanolaena maxima	12.27		13.08
29.	Vetiveria zyzanioides	29.07		31.342

TABLE 7.14
Periodic Estimation of Binding Capacity of Selected Species in Natural Habitat, Tailing Ponds and in Experimental Containers

Sl.No.	Plant species	Natural habitat	Tailing pond		Experimental containers	
			After 1 Yr	After II Yr	After 1 Yr	After II Yr
1.	*Colebrookea oppositifolia*	97.9	69.44	70.44	82.64	84.44
2.	*Dodonaea viscosa*	174.0	105.0	106.0	109.56	110.22
3.	*Furcaria foetida*	160.0	55.0	55.0	92.45	*92.94*
4.	*Imperata cylindrica*	111.1	99.44	100.0	102.56	104.0
5.	*Jatropha gossypifolia*	*104.0*	102.0	102.6	111.11	112.64
6.	*Pogostemon benghalense*	104.9	104.93	105.0	117.64	118.44
7.	*Saccharum spontaneum*	88.0	85.0	86.0	86.0	87.50

TABLE 7.15
Binding Capacity (Conservation Value) at Experimental Sites

S. No.	Plant species	Conservation value		
		Ex situ	In situ	Natural forest
1.	*Alysicarpus monilifer*	52.147	52.577	70.85
2.	*Argemone mexicana*	6.55	10.99	10.738
3.	*Bothriochloa intermedia*	73.93	0	74.407
4.	*Bothriochloa pertusa*	66.595	0	73.783
5.	*Cenchrus ciliaris*	98.785	61.553	72.773
6.	*Cenchrus setigerus*	55.692	39.83	64.33
7.	*Chrysopogon fulvus*	62.41	0	71.84
8.	*Cleome viscosa*	44.29	56.362	56.96
9.	*Clitoria ternatea*	0	40.48	30.76
10.	*Crotalaria sericea*	21.205	19.471	16.70
11.	*Crotalaria juncea*	34.17	0	23.49
12.	*Croton bonplandianus*	15.96	12.30	70.35
13.	*Cymbopogon citratus*	0	39.43	43.94
14.	*Desmanthus virgatus*	36.40	0	39.45
15.	*Desmodium gangeticum*	33.47	24.11	33.61
16.	*Dicanthium annulatum*	65.58	56.18	67.97
17.	*Heteropogon contortus*	47.40	57.65	57.192
18.	*Hyptis suaveolens*	0	38.56	37.85
19.	*Ingigofera trita*	0	27.40	29.03
20.	*Iseilma laxum*	0	46.71	47.83
21.	*Mucuna pruriens*	24.3	0	29.40

S. No.	Plant species	Conservation value		
		Ex situ	*In situ*	Natural forest
22.	*Ocimum canum*	0	33.88	33.88
23.	*Panicum antidotale*	103.26	0	124.49
24.	*Pennesetum caudatum*	46.37	0	47.605
25.	*Rhynchosia minima*	0	38.18	46.55
26.	*Stylosanthes hamata*	31.49	0	33.04
27.	*Tephrosia purpurea*	0	19.40	20.79
28.	*Thysanolaena maxima*	46.817	0	51.334
29.	*Vetiveria zyzanioides*	72.109	0	125.9

Binding capacity of selected species was evaluated both under field conditions and in experimental containers. *Dodonaea viscosa* has the maximum value of binding capacity followed by *Pogostemon benghalense, Imperata cylindrica and Colebrookea oppositifolia. Saccharum spontaneum* has the lowest binding capacity.

7.2.4 COMPARATIVE STUDY OF ROOT PENETRATION OF SELECTED SPECIES IN EXPERIMENTAL CONTAINERS (*EX SITU*), TAILING POND (*IN SITU*) AND NATURAL HABITAT

Root penetration (cm) studies showed variations among the three sites of experimentations *viz. in situ, ex situ* experimental site and natural habitat, and among the seven species. In short, all these factors showed significant variations in respect of root penetration.

Root penetrations (cm) in the soil among the seven species have been found to vary significantly. In *ex situ* experimental trials, it was found that species *Pogostemon benghalense* (22 cm) contains the maximum amount of root penetration in its roots while the species *Saccharum spontaneum* (15 cm) was found to contain the least amount of root penetration capacity among the seven species. In *in situ* experimental trials, it was found that species *Pogostemon benghalense* (22.20 cm) contains the maximum root penetration capacity in its roots while the species *Saccharum spontaneum* (15.00 cm) was found to contain the least amount of root penetration capacity among the seven species. In natural habitat, it was found that species *Pogostemon benghalense* (24.50 cm) contains the maximum amount of root penetration capacity in its roots while the species *Furcraea foetida* (16.00) was found to contain the least amount of root penetration capacity among the seven species. Root penetrations of the roots among the seven species have been found nearly the same in *ex situ* and *in situ* experimental trials, but in natural habitat it was statistically significant. In natural habitat, soil root penetration was much greater than *ex situ* and *in situ* experimental trials.

The comparative root length of plant species on a revegetated tailing pond site (*in situ*) and experimental containers site (*ex situ*) were determined. The length of roots varied in both species and as well as between the two contrasting sites. *Pogostemon benghalense* proved to be superior in terms of its contribution to total root length on revegetated site as well as natural habitat. The extensive growth of this species was undoubtedly due to the presence of an extensive network of fibrous roots. It can therefore be concluded that all seven species were effective soil binders and suitable soil conserving species for tailing pond areas. Overall results reveal that all the selected species were good soil binders vis-à-vis being fast growing in nature. It was therefore recommended that these species should be given priority to plant on degraded lands for soil conservation and revegetation purposes.

Soil binding capacity of the roots among the seven species (in Phase I) varied significantly. In *ex situ* experimental trials, it was found that the species *Pogostemon benghalense* (118.04) contains the maximum amount of soil binding capacity in its roots, while the species *Colebrookea oppositifolia* (83.54) was found to contain the least amount of soil binding capacity among the seven species. In *in situ* experimental trials, it was found that the species *Dodonaea viscosa* (105.50) contains the maximum amount of soil binding capacity in its roots, while the species *Furcraea foetida* (55.00) was found to contain the least amount of soil binding capacity among the seven species. In natural habitat, it was found that the species *Dodonaea viscosa* (174.00) contains the maximum amount of soil binding capacity in its roots, while the species *Saccharum spontaneum* (88.00) was found to contain the least amount of soil binding capacity among the seven species (Table 7.14). Soil binding capacities of the roots among the seven species have been found nearly same in *ex situ* and *in situ* experimental trials but in natural habitat it was statistically significant. In natural habitat, soil binding capacity was much greater than in *ex situ* and *in situ* experimental trials.

The analysis of data of binding capacity showed variation between the two age group data, between the two sites of experimentations *viz. in situ* and *ex situ* experimental sites, and among the seven species. In short, all these factors showed significant variations in respect of binding capacity (conservation value). It can be observed that binding capacity in the secondage group plants was more (95.30) than the first age group plants (94.48). Analysis showed that this difference was significant. Hence, it may be explained that binding capacity has increased.

The average binding capacity in experimental containers was 100.80, which was significantly more than that of tailing pond where the binding capacity was 88.99. As mentioned that species vary significantly, it was observed that soils under the species *Pogostemon benghalense* showed highest binding capacity (111.50), whereas the species *Furcraea foetida* showed least binding capacity (73.85) among the seven species. Species *Dodonaea viscosa* and *Jatropha gossypifolia* showed equal binding capacity (107.70 and 107.09). The age-wise averages of sites showed that binding capacity of experimental containers remains higher than the tailing pond in second age group plants. In comparing the species, it was observed that the binding capacity of all the species improves in the second age group plants. The species *Pogostemon benghalense* remains higher than and *Furcarea foetida* remains lowest of other species in both age groups of plant.

When site-wise binding capacity of species was compared, it was observed that species perform significantly better in experimental containers than in the tailing pond, that is, the binding capacities of all the species in experimental containers were far greater than that observed in the tailing pond. Viewing the joint effect of the three factors on binding capacity, it may be observed that all the species in experimental containers show higher binding capacities, and showed further increase in the second age group. Tailing pond in both age groups showed lower binding capacity for all the species. Root volume of selected species on tailing pond for two consecutive years showed that there was a significant annual increase in the root biomass and volumes of selected species. Another important observation made on the tailing pond is that shoot growth in terms of height of species appeared to be low because of trampling by cattle.

7.3 RADIONUCLIDE STUDIES IN PLANTS, SOIL AND TAILINGS

The mobility of weathered uranium in the soil profile is dependent upon the affinity of the soil for uranium and properties of the soil, such as its acidity or alkalinity and water content. Thus, where soil has a high affinity for binding uranium and its release into soil (typically soils high in organic matter),water should be minimal. Correspondingly, mobility is likely to be greater in soils that bind uranium less strongly, which includes soils of semiarid climate where neutral to alkaline soil pH is combined with low organic carbon content (Ebbs et al., 1998). Plant uranium concentration was mostly affected by concentration of uranium in the soil.

Study was undertaken to consolidate the radioactivity in the tailings using native primary colonizing plant species. Under this program, a tailing pond was covered with a 30 cm layer of soil. Seeds of the selected plant species were sown over the soil-covered tailing pond for the study of radionuclide uptake (*in situ* trials). An attempt was also made to grow the same species under *ex situ* conditions in 200-liter experimental containers in Jaduguda. These experimental containers were filled with tailings and then covered with 30 cm soil as in the field trials and then planted. These experimental trials have been conducted to compare the binding capacity and root penetration of selected species in both *ex situ* and *in situ* conditions. Species selection for these experimental trials was made after surveying five tribal villages surrounding the tailing pond to ascertain the nonusability of the species for edible or other purposes. The other important criteria for selection are their growth form *viz.*, perennials with low crown cover, shallow root system and their conservation value. Seven species have been selected and tried for *in situ* as well as *ex situ* experiments. Plant samples both above and below ground, soils and tailing samples were collected from different depths from the study sites marked as tailing pond (*in situ* experiment) and experimental containers (*ex situ* experiment located at Jaduguda). Soil samples were also collected from village land and forest land on the periphery of the tailing pond for comparing the physical and chemical properties of soils. Sampling was done on an annual basis. The samples collected were analyzed and the mean values of two years were taken. Natural uranium, polonium and radium were recorded.

7.3.1 RADIOACTIVITY STUDIES IN PLANTS

Concentrations of uranium (natural) in different parts of the selected plant species were evaluated. Soil and tailing samples at different depths adhering to the roots of the plants were also analyzed to estimate the concentration ratio. In certain cases for radiochemical analysis, entire aboveground parts were analyzed and reported as root/shoot. Analytical results of vegetation samples were reported on a dry weight basis. The order of selectivity was much more for root, and the least accumulation was observed in the shoot. Variability in the root/soil uranium concentration ratio observed was probably due to the presence of small soil particles (observed by microscopic examination), presumably uranium adhering to the roots. It was further speculated that uranium colloids may have been sorbed on the root surface. This further complicates the differentiations of "in" vs. "on" uranium components in plant roots.

Results indicated that the amounts of uranium in roots among the seven species have been found to vary significantly. It was found that *Pogostemon benghalense* contains the maximum amount of uranium in its roots (37.55 Bq/kg) in dry weight of root, while *Jatropha gossypifolia* was found to contain the least amount of uranium (2.691 Bq/kg) among the seven species. The total amount of uranium in roots is in the order *Pogostemon benghalense* (37.55 Bq/kg) >*Imperata cylindrica* (32.338 Bq/kg) >*Colebrookea oppositifolia* (30.944 Bq/kg) >*Saccharum spontaneum* (15.637 Bq/kg) >*Dodonaea viscosa* (8.731 Bq/kg) >*Furcraea foetida* (4.777 Bq/kg) >*Jatropha gossypifolia* (2.691 Bq/kg). According to Jha et al. (2001),a relatively lower concentration profile was observed in *Ipomoea carnea* (amri). Concentration in roots varied from 2.25–80.0 Bq/kg with geometric mean and arithmetic mean 11.5 and 19.0 Bq/kg and geometric standard deviation and standard deviation 69.25 and 24.25 Bq/kg. In certain cases, concentration in stem and leaf was comparable, whereas in other cases it may even exceed that of the concentration in root. This exceptional characteristic of higher concentration in leaf as compared to root was found only in this species of tailing pond. In the entire set of plants analyzed from the tailing pond, root always had a concentration greater than that in stem or leaf. Concentration in stem varied from 0.675–9.0 Bq/kg with geometric mean and arithmetic mean 1.0 and 2.7 Bq/kg and geometric standard deviation and standard deviation 152 and 3.5, whereas it ranged from 1.25 to 77.25 Bq/kg in leaf with geometric mean and arithmetic mean 10.0 and 22.75 Bq/kg and geometric standard deviation and standard deviation 105.75and 28.25. In four out of nine samples analyzed from different locations of the tailing pond, leaf was showing a concentration greater than that in root Jha et al. (2001). Entire aboveground parts were analyzed, and the concentration in root was significantly greater than that in aboveground parts (384.75 and 22.5 Bq/kg). In *Cyperus rotundus*, the highest concentrations among all the samples analyzed were found in both roots and aboveground parts of the species. Concentration in root was 3172.5 Bq/kg and in aboveground parts it was 317.25 Bq/kg. The plant may serve as an indicator owing to its quite high preferential uptake. Concentration of uranium in *Croton bonplandianum* leaf and root was comparable, whereas concentration in stem and branch was only slightly lower (root: 11.5, stem and branch 8.25 and leaf 13.75 Bq/kg) (Jha et al., 2001). The higher concentration ratios for uranium on the mill tailings would suggest that there is some dependence of uptake of these radionuclides on initial soil concentration. The long-term implication of radionuclide uptake involves the accumulation of uranium on the soil surface from many years of litter fall and decomposition, as well as off-site movement by wind and water. Study showed that concentration of root uranium was significantly correlated with shoot uranium, but it showed a negative correlation with shoot radium percentage. The results were comparable with Jha et al. (2007).

The tried species were found to vary significantly in respect to uranium in their shoots. It was found that *Saccharum spontaneum* possesses the maximum amount of uranium in its shoots (9.356 Bq/kg), while *Jatropha gossypifolia* contains the least amount of uranium (1.410 Bq/kg) in its shoots. It is further seen that *Saccharum spontaneum* contains 6.64 times more uranium than that of *Jatropha gossypifolia*. Jha et al. (2001) reported that *Lantana camera* (putus), which was collected at the edge of the tailing pond, showed a concentration of uranium in its root varying from 3.9 to 78.0 Bq/kg, whereas in the shoot it varied from 5.0 to 18.0 Bq/kg. Apart from this, analysis of other less abundant species not normally found in the concerned environment were also carried out, including *Ageratum conyzoides*, which was derived from the nearby hillside where no artificial activity was taking place. The natural weathering process enhances the sorption capability of substrate due to accumulation of weathered organic matter. More equal distribution of uranium across stem and leaves may lead to a slower return of uranium in the litter and/or long-term sequestration of uranium in woody debris. The age-wise performance of the species indicates that the uranium in all the species showed an increase with age. The combination values differed significantly. The overview of three-dimensional results leads to a similar conclusion mentioned earlier. The shoots of *Saccharum spontaneum* contain the highest level of uranium in both the sites during both years among the seven species.

Percentage of uranium in the shoot indicated that uranium concentration varied in different species. Uranium percentage in shoots was significantly higher in experimental containers (46.42%) than in the tailing pond (29.27%). The age-wise data show that uranium percent in shoot increased with age. The data are comparable to the data from Hossner et al. (1997). Regarding the amounts of polonium in roots, *Saccharum spontaneum* contained the maximum amount of polonium (36.62 Bq/kg), while *Jatropha gossypifolia* was found to contain the least amount of polonium (4.37 Bq/kg) among the experimental species. Studies by Jha et al. (2001) showed that the concentration of polonium was higher in roots and shoots as compared to present studies. The species differed significantly among themselves in respect of polonium content in shoots. *Saccharum spontaneum* possesses the highest amount of polonium (20.064 Bq/kg). *Jatropha gossypifolia* lags behind all the species as it was found to contain the lowest amount of polonium (3.55 Bq/kg). From the interaction results of age and site, it may be observed that polonium in experimental containers as well as tailing pond area showed no significant change in second age. The polonium in all the species showed a small but insignificant increase in the second age. Jha et al. (2007) recorded that polonium concentration in native vegetation of the Banduhurang uranium mine site was higher in comparison to the present study in the uranium tailing pond area. The concentrations of *Vitex negundo* roots (30.53 Bq/kg) and shoots (16.67 Bq/kg), *Cassia tora* roots (11.11Bq/kg) and shoots (14.65 Bq/kg), and *Sida cordifolia* roots (42.50 Bq/kg) and shoots (11.38 Bq/kg) were recorded by Jha et al. (2007). Site-wise performance of species in respect of polonium content in the shoots showed that most of the species contain a higher amount of polonium in the tailing pond. It was found that the percent of polonium varied highly significantly among the species, and *Saccharum spontaneum* contains the highest percentage (35.524%) of polonium while *Jatropha gossypifolia* (4.998%) contains the lowest percentage. It may be therefore concluded that *Saccharum spontaneum* was more prone to absorption of polonium than other species, whereas *Jatropha gossypifolia* had poor absorption ability. The age-wise average showed that, in general, percent intake of polonium in the experimental areas by the species increased with age.

Radium found in roots reveals that there was significant variation between sites, ages and species. It was observed that the amount of radium was significantly (30.68%) higher in *in situ* (tailing pond) experimental sites (1.07 Bq/kg) than in *ex situ* (experimental containers) (0.82 Bq/kg). The age-wise averages showed that the level of radium in roots increased significantly (5.69%) in the second age of experimentation (0.97 Bq/Kg) in comparison to first age level (0.92 Bq/kg) (Table 7.33). Tsivoglue (1964) used filamentous algae in monitoring Animas River radioactivity in the vicinity of uranium mills in Colorado, USA. ^{226}Ra concentration of 148 Bq/kg of algal ash were recorded for uncontaminated streams, whereas a concentration as high as 129.5×103 Bq/kg of algal ash was recorded for the polluted locations of the Animas River, which showed a concentration potential of algae of a high order for ^{226}Ra. A new trend has developed to carry out environmental impact assessment following uranium mill effluent discharge through the study of radioactivity associated with the algal bodies growing in the habitat. Markose et al. (1981) observed an elevated concentration of radium in algae from Jaduguda, India. It was found that the variation in concentration is dependent on the proximity of location with respect to tailing pond effluent or mine water streams, which are a potential source of ^{226}Ra in streams. The concentration factor ranged from 800 to 3100, which is in agreement with those (500–1000) reported by Tsivoglue (1964). ^{226}Ra level in near-shore marine plankton have been studied at Kalpakkam, and the concentration was found to be 1.32 and 1.55 Bq/kg wet, with concentration factors of 825 and 968 in composite samples of phytoplankton and zooplankton respectively (IAEA, 1987). However, the ^{226}Ra concentration factor for freshwater plankton was reported to be 2775–6820, which is much higher than that reported for the near-shore environment of Kalpakkam (2.78 Bq/g wet in plankton and in water 0.4 Bq/l–1.0 Bq/l). Investigations on the uptake of radium by terrestrial plants demonstrate that some plants may serve as indicators of radium in soil. Quantitatively, more radium may be accumulated by plants grown on tailings; however, the uptake of radium is a function of available radium in the soil matrix and not dependent on total radium. Tailings were found to contain less percentage of available radium compared to normal soil. *Abelmoschus esculantus* (lady's finger) accumulated more radium, and close relationship was found between Ra and Ca. Roots and leaves accumulate more radium, while least accumulation was observed in fruits (Markose, 1990).

The amounts of radium in root among the seven species have been found to vary significantly. It was found that species *Furcraea foetida* possesses the least amount of radium in its roots (0.15 Bq/kg), while the species *Imperata cylindrica* contains the maximum radium concentration in its roots (1.02 Bq/kg) among the seven species. The joint effects of ages and sites show that radium in both sites have increased in the second age as compared to the first age. During both ages, radium in the roots of selected species *in situ* remains higher than *ex situ*. It conforms to the overall findings of the sites, i.e., *in situ>ex situ* (Table 7.37). The present study is comparable with other past study in the uranium mill tailings of South Dakota, the mean value of ^{226}Ra is estimated to be 4107 Bq/kg dry in contrast to the controlled areas, where it was found to be 66.6 Bq/kg dry sample (Rumble and Bjugstad, 1986).

Concerning site-wise performance of species in respect of radium in the roots, it can be inferred that all the species contain a higher percentage of radium in the tailing pond (1.0754 Bq/kg) area. The interaction values differ highly significantly. An over-all view of the effect of the three factors leads to the same conclusion as discussed earlier. The species *Furcraea foetida* showed higher values of radium and *Imperata cylindrica* the lowest values as compared to other species in both ages in both sites. All the species show a rise in radium level after one age group, i.e., in the second age group. The species possesses higher values of radium in the *in situ* (tailing pond) experimental site (Table 7.38). Uptake of ^{226}Ra by water lily (*Nymphaea* sp.) was investigated by Martin and Ryan (2004). It was found that concentrations of ^{226}Ra in the rhizome are higher than in foliage, this being due to surface accumulation. Uptake of radium by the foliage was found to be primarily from the water rather than by translocation from roots. Analysis of the ratio of radium and calcium in plants, compared to extractable concentration in water and sediment, showed no correlation between foliage and supporting media, suggesting that different mechanisms were involved in accumulation of radium and calcium. Davy et al., (1974) analyzed fruit samples of apple (*Engenia* spp.) and fig (*Ficus henneana*) for radium content grow-ing near the Alligato River area of Australia. The values obtained ranged between 0.15–15 Bq/kg fresh and 1.1–1.8 Bq/kg fresh respectively. As per their observation, there was no significant difference in the specific activity of radionuclides present in the food samples. Martin and Ryan (2004) reported activity concentrations as well as concentration ratios relative to total, soil activity for samples of cheeky yam (*Dioscorea bulbifera*) and anbulubi roots (*Eriosema chinense*). Difference between the activity concentrations in two species (0.26 and 1.19 Bq/kg fresh) was partly attributed to the peeling of yam and the non peeling of anbulubi. Jha et al. (2001) and Khan et al. (2000) have evaluated maximum concentration of radium in selective species. There exists significant difference in radium between the two sites. Radium was significantly higher (63%) in the tailing pond (0.225 Bq/kg) than in experimental containers (0.138 Bq/kg). The averages of the ages showed that the amount of radium decreased in the second age of experimentation as compared to the first age, although the difference was statistically nonsignificant. It was 0.193 Bq/kg in the first age while in the second age it reduced to the level of 0.170 Bq/kg. The interaction values recorded vary significantly. This is in contrast to polonium and uranium, where a significant increase has been recorded in the second age. The data of present studies are comparable with Jha et al. (2001) and Khan et al. (2000), who reported maximum uptake of radium by selective species in an open tailing pond area.

The amounts of radium in shoots of different species varied significantly. *Saccharum spontaneum* contained the maximum amount of radium in its shoot (0.388 Bq/kg), while *Jatropha gossypifolia* contains the least amount of radium (0.045 Bq/kg) among all the tried species. It was observed that *Saccharum spontaneum* holds 8.62 times more radium as compared to the *Jatropha gossypifolia*. However, results indicate that all the species have a higher amount of radium in the tailing pond as compared to that of experimental containers. The values varied highly significantly. According to Jha et al. (2001),*Cyperus rotundus* showed concentration in the root and aboveground parts as 392.2 Bq/kg and 71 Bq/kg respectively, whereas in *Croton bonplandianum*, concentration in root, stem and leaf was 7.9, 12.3 and 31.1 Bq/kg respectively. In *Croton bonpladianum*, the migration reflects poor retention in the root as compared to aboveground parts.

Further, concerning the uptake and accumulation of radium in the seven species, it was observed that *Jatropha gossypifolia* and *Furcraea foetida* have the lowest uptake and accumulation of radium, while *Saccharum spontaneum* and *Pogostemon benghalense* have the highest accumulation of radionuclide substances. *Jatropha gossypifolia* and *Furcraea foetida* have the lowest uptake and accumulation of radionuclides, while *Saccharum spontaneum* and *Pogostemon benghalense* have highest accumulation of radionuclide substances. Works carried out by several workers lead one to infer that different parts of the same plant show marked differences in concentrations of elements, which probably change with time. This has been observed for nutrient elements as well as nonessential elements.

Radium percentages in shoots among the studied species varied significantly. It was found that the species *Furcraea foetida* along with *Jatropha gossypifolia* possess the maximum amount of radium percent in their shoots (34.3% and 35.6%), while *Imperata cylindrica* contains the least radium (15.2%) among the seven species. It may be recalled that species *Imperata cylindrica* contains the maximum percent of polonium (74.77%).Radium content of native and neutralized plant species growing at inactive uranium mill sites was shown to be significantly greater than that of the local plants. This was shown on both covered and uncovered tailings and on native soils with elevated radium content. Similar values were found in two species grown on a soil-covered tailing pond. Thus, radium is adsorbed and translocated by plants, and this could be a significant mechanism of dispersal in the environment. Radium is a contaminant in all uranium mill tailings unless removed by leaching with nitric acid. Even if tailings are buried beneath deep soil overburdens, biological penetration of the cover by plant root of deep-rooted plant species such as *Atriplex canescens* and burrowing animals could introduce radium into the food chain.

7.4 NATURAL URANIUM IN SOIL AND TAILINGS

Uranium concentration in soil (0–15, 16–30 cm) and tailing (30–45 cm) samples collected from two sites/experimental areas from three depths and for two consecutive years was analyzed as factorial analysis using statistical packGenStat-32 version 5. The soil uranium differs significantly with depths. The depth of the soil was an influencing factor for the presence of uranium. It was observed that uranium increases with increasing depth. Its quantity at the upper surface upto 0–15 and 15–30cm was very low (86.25 Bq/kg, 108.75 Bq/kg). However, it seems to be in abundance in tailings beyond 45 cm (826.25 Bq/kg). The analysis of uranium concentration in soil and tailings does not show any significant difference between the sites. The soil uranium in both sites may be termed as statistically the same. It may be asserted that soil uranium of both experimental sites was homogenous (Table 7.16).

TABLE 7.16

Concentration of Uranium (Bq/kg) of Soil and Tailings from Experimental Containers and Tailing Pond

Year	Site	Depth (cm) Soil 0–15	Soil 15–30	Tailing 30–45	Mean
	Exp. containers	62.50	77.50	65.00	**67.50**
1st Year	Tailing pond	112.50	140.00	1080.00	**445.00**
	Mean	*87.50**	*110.00*	*572.50**	*255.00*
	Exp. containers	60.00	77.50	1092.50	**410.00**
2nd Year	Tailing pond	110.00	140.00	1067.50	**440.00**
	Mean	*85.00*	*107.00***	*1080.00***	*425.00****
Mean of depth		**85.25**	**108.75**	**826.25**	**340.625**
Mean of Exp. Site		Exp. Containers **240.00**		Tailing Pond **442.50**	

CD	Depth	Exp. Site	Year	Dept*Exp. Site
	263.25	215	215	372.25
	Depth*Year	Exp. Site*Year	Depth*Exp. Site*Year	
	372.25	304	526.5	

Note:* Significance at 5% level **Significance at 1% level *** Significance at 0.01% level

It was also revealed that factor-depths have a marked influence on soil uranium. It was observed that the variation among three depths was highly significant. It was noted that the value of soil uranium increases with the depth of the soil in both sites. Its quantity in deeper soil, i.e., 30–45 cm, was found to be nearly ten times more than that of the uppermost soils, i.e., up to the depth of 30 cm. Uranium up to the depths of 30 cm was low in quantity and consistent, but beyond 30 cm it is high (which is tailings). No significant change in soil uranium values over the span of two years has been observed, which means that the soil in both experimental trials over the period of two years has remained the same in respect of soil uranium levels. The data are comparable with Jha et al. (2001), Khan et al. (2000) and Jha et al. (2005a); the concentration of uranium in the tailing pond area was recorded at 193 Bq/kg to 3200 Bq/kg before covering the soil.

Sarangi et al., (2009) has explained that in the whole Nagpur plateau, the concentration of uranium of 0.5to 4 ppm at 0–50 cm depth means 12.5 Bq/kg to 100 Bq/kg was recorded. After covering the soil in the tailing pond area, the concentration of uranium recorded was the same, according to Sarangi et al., (2009). Soil uranium differs significantly with depths. The depth of the soils was an influencing factor for the presence of uranium. It was observed that uranium increases with increasing depth. Its quantity at the upper surface upto 30 cm was very low. It seemed to be in abundance in soil beyond 45 cm.

Soil (0–15, 15–30 cm) and tailing (30–45 cm) samples collected from two sites/experimental areas from three depths and for the two consecutive years showed that the depth of the soil was an influencing factor for the concentration of uranium.

It was observed that uranium increases with an increase in the depth of soil. Its quantity at the upper surface upto 0–15 and 15–30cm was very low (86.25 Bq kg^{-1}, 108.75 Bq kg^{-1}), but after this depth it significantly increases. Its quantity in tailings, i.e., 30–45 cm (826.25 Bq kg^{-1}) depth was nearly ten times more than that of the upper strata of soils, i.e., up to the depth of 30 cm. No significant change in soil uranium concentrations over the span of two years has been observed, which means that the soils in both experimental trials over the period of two years have remained same in respect of soil uranium levels. The estimation of uranium concentration in soil and tailings does not show any significant difference between the sites. The uranium concentration in both sites was statistically the same (Table 7.17).

Data from four sites and three depths showed that soil uranium varied with sites and depths. The soil uranium was found to be highest in soil of *in situ* experimental trials (tailing pond) and closely followed by *ex situ* experimental soil (experimental containers). The soiluranium of the rest of the sites was low in levels and statistically the same. It was observed that the uranium in the soils of the tailing pond and the experimental containers was nearly 17 times more than that of forest land and village land soil. It was observed that uranium increases with increasing depth. Its quantity at the upper surface upto 30 cm was very low. It seemed to be in abundance in soil beyond 45 cm (Jha et al., 2005b).

Soil uranium in the soil samples from four sites and three depths showed that the soil uranium varied with sites and depths. Soil uranium was found to be highest in soil of *in situ* experimental trials, closely followed by *ex situ* experimental soils. The soil uranium of the rest of the sites was low in levels and statistically the same. It was observed

TABLE 7.17

Concentration of Uranium (Bq/kg) of Soil and Tailings from Experimental Containers and Tailing Pond

Year	Site	Depth (cm)			Mean
		Soil		Tailing	
		0–15	15–30	30–45	
1st Year	Exp. containers	66.0	77.50	65.00	**67.50**
	Tailing pond	64.0	140.00	1080.00	**445.00**
	Mean	*87.50**	*110.00*	*572.50**	*255.00*
2nd Year	Exp. containers	60.00	77.50	1092.50	**410.00**
	Tailing pond	110.00	140.00	1067.50	**440.00**
	Mean	*85.00*	*107.00***	*1080.00***	*425.00****
Mean of depth		**86.25**	**108.75**	**826.25**	**340.625**
Mean of Exp. Site		Exp. Containers **240.00**		Tailing Pond **442.50**	
CD	Depth	Exp. Site	Year	Dept*Exp. Site	
	263.25	215	215	372.25	
	Depth*Year	Exp. Site*Year	Depth*Exp. Site*Year		
	372.25	304	526.5		

Note: * Significance at 5% level ** Significance at 1% level *** Significance at 0.01% level

that the uranium in the soils of the tailing pond and the experimental containers was nearly 17 times more than that of forest land and village land soils (Table 7.18).

It was observed that uranium increases with increasing depth. Its quantity at the upper surface upto 30 cm was very low. It seemed to be in abundance in soil beyond 45 cm. The data pertaining to uranium concentration in soil (0–15, 15–30 cm) and tailing (30–45 cm) samples collected from two sites/experimental areas from three depths and for the two consecutive years. Factorial analysis of this data was done using statistical packGenStat-32 version 5. It was observed that uranium increases with increase in depths of the soils. Its quantity at the upper surface upto 0–15 and 15–30 cm was very low (86.25 Bq/kg, 108.75 Bq/kg) but after this depth it significantly increases. Its quantity in tailings i.e., 30–45 cm (826.25 Bq/kg) depth was nearly ten times more than that of the upper strata of soils, i.e., up to the depth of 30 cm. No significant change in soil uranium concentrations over the span of two years has been observed, which means that the soils in both experimental trials over the period of two years has remained same in respect of soil uranium levels. The estimation of uranium concentration in soil and tailings does not show any significant difference between the sites. The uranium concentration in both sites was statistically the same.

Results further reveal that factor-depths have marked influence on the soil uranium. The variation among three depths was highly significant. It was noted that concentration of uranium increases with the depth of the soils in both sites. The analysis of soil uranium in the soil samples from four sites and three depths showed that the soil uranium varied with sites and depths. Uranium in the soils of tailing pond and the experimental containers was nearly 17 times more than that of adjoining forest and village land soils. It was observed that uranium increases with increasing depth. Its quantity at the upper surface upto 30 cm was very low. It seemed to be in abundance in soil beyond 45 cm.

TABLE 7.18
Results of Analysis of Concentration of Uranium (Bq/kg) of Soil and Tailings in Different Experimental Sites

Site	Depth (cm)			Mean
	0–15	15–30	30–45	
Tailing pond (*in situ*)	112.50	140.00	1080.00•	445.00
Village land	32.50	37.50	42.50	37.50
Forest land	35.00	37.50	42.50	37.50
Exp. Containers (*ex situ*)	62.50**	77.50**	1102.50•	412.50
Mean	**60.00**	**72.50**	**567.50**	**232.50**
CD	Depth	Exp. Site	Depth* Exp. Site	
	215.75	249.25	431.50***	

Note: * Significance at 5% level ** Significance at 1% level *** Significance at 0.01% level
• Tailings

7.5 POLONIUM IN SOIL AND TAILINGS

Polonium concentration did not show any significant difference between the sites. However, depths showed a marked relationship with the soil polonium. It was observed that the variation among the three depths was highly significant—the soil polonium concentration increases with the depth of the soil in both sites. Concentration of polonium in tailings, i.e., 30–45 cm, was nearly five times more (329 Bq kg^{-1}) than that of the surface soil layer (65.0 Bq kg^{-1}), i.e., up to the depth of 15 cm. It was also noted that variation in soil polonium values over the span of two years showed no significant difference (Table 7.18). According to Jha et al. (2001), polonium concentration on the surface of the uranium tailings (0–10cm) was 2200 Bq kg^{-1}. In the present study after covering with the soil (30 cm), concentration of polonium was very low (20–65 Bq kg^{-1}).

The analysis of concentration of polonium showed that its behavior was very similar to that of uranium. The polonium in the soils collected from four sites and three depths showed that the content of polonium in the soil varies with sites and depths as well. The concentration was found to be highest in experimental containers' soil, closely followed by soil of the tailing pond. Statistically, soils of experimental containers and tailing ponds were similar with respect to polonium content. As already mentioned, the quantity of polonium varied significantly with depths. It has been observed that it showed an increasing trend with increasing depth. The difference between depth 0–15 (42.3 Bq kg^{-1}) and 15–30 (58.9 Bq kg^{-1}) was not significant; it was the third depth 30–45 cm, which is actually tailings (176.10 Bq kg^{-1}), where polonium seems to be in abundance and significantly higher than the first two depths, as mentioned earlier (Table 7.19). According to Jha et al. (2007), concentration of polonium in the soil of Banduhurang uranium mine areas were recorded at 186.10 to 1200 Bq kg^{-1}. The analysis of soil polonium showed that it has positive correlation with soil organic carbon and phosphorus. Polonium concentration did not show any significant difference between the sites. However, depths showed a marked relationship with the concentration of polonium in the soils.

The variation among three depths was observed as highly significant, that the polonium concentration increases with the depth of the soil in both sites. Concentration of polonium in tailings, i.e., 30–45 cm, was nearly five times more (329 Bq kg^{-1}) than that of the surface soil layer (65.0 Bq kg^{-1}), i.e., up to the depth of 15 cm. It was also noted that variation in soil polonium values over the span of two years showed no significant difference.

The analysis of quantity of polonium showed that its behavior was very similar to that of uranium. The polonium in the soils collected from four sites and three depths showed that the content of polonium in the soils varied with sites and depths as well. It was found to be highest in quantity in experimental containers' soils, closely followed by soils of the tailing pond. Statistically, soils of experimental containers and the tailing pond were the same in polonium content. As already mentioned, the quantity of polonium varied significantly with depths. It has been observed that it showed an increasing trend with increasing depth. The difference between depths 0–15 cm and 15–30 cm was not significant; it was the third depth where polonium seems to be in abundance and significantly higher than the first two depths, as mentioned earlier.

In the case study undertaken at Jaduguda, polonium did not show any significant difference between the experimental sites. However, depths showed a marked relationship

TABLE 7.19

Concentration of Polonium (Bq/kg) of Soil and Tailings in Both *Ex Situ* and *In Situ* Experimental Trials

Year	Site	Depth (cm)			Mean
		Soil		Tailing	
		0–15	15–30	30–45	
1st Year	Exp. containers	66.0	81.0	366.0	171.0
	Tailing pond	64.0	112.0	295.0	157.0
	Mean	*65.0**	*96.0*	*331.0**	*164.0*
2nd Year	Exp. containers	66.0	80.0	364.0	170.0
	Tailing pond	64.0	112.0	294.0	156.0
	Mean	*65.0*	*96.0***	*329.0*	*163.0*
Mean of depth		65.0	96.0	330.0	
Mean of Exp. Site		Exp. Containers **171.00**		Tailing Pond **157.0**	
CD	Depth	Exp. Site	Year	Dept*Exp. Site	
	53.7	43.9	62.0	76.0***	
	Depth*Year	Exp. Site*Year	Depth*Exp. Site*Year		
	76	62.0	526.5	107.4	

Note: * Significance at 5% level **Significance at 1% level *** Significance at 0.01% level •Tailings

with the concentration of polonium in the soils. It was observed that the variation among three depths was highly significant, that the soils' polonium concentration increases with the depth of the soil in both sites. Concentration of polonium in tailings, i.e., 30–45 cm, was nearly five times more (329 Bq/kg) than that of the surface soil layer (65.0 Bq/kg), i.e., up to the depth of 15 cm. It was also noted that variation in soil polonium values over the span of two years showed no significant difference (Table 7.20).

TABLE 7.20

Concentration of Polonium (Bq/kg) of Soil and Tailings in Different Experimental Sites

Site	Depth (cm)			Mean
	0–15	15–30	30–45	
Tailing pond (*in situ*)	64.0	111.24	295.5*	**157.3**
Village land	17.3	18.1	18.2	**17.9**
Forest land	22.1	24.4	24.6	**23.7**
Exp. Containers (*ex situ*)	65.8	80.6**	366.2*	**170.9**
Mean	**42.3**	**58.9**	**176.1**	**92.4**
CD	Depth	Exp. Site	Depth* Exp. Site	
	38.40	44.34	76.79***	

Note: * Significance at 5% level **Significance at 1% level *** Significance at 0.01% level •Tailings

7.6 RADIUM IN SOIL AND TAILINGS

The estimation of radium concentration in soil showed that its concentration was very similar to the other radionuclides, uranium and polonium. It was also found that depth was an influencing factor for the concentration of radium. It was observed that the variation among three depths was highly significant. It was noted that the value of soil radium increases with the depth of the soil in both sites. Its quantity in deeper layers, i.e., 30–45 cm (tailings), was found to be nearly four times more (34.25 Bq kg^{-1}) than that of the upper soil layers (8.20 Bq kg^{-1}), i.e., up to 15 cm. It was also observed that variation in soil radium values over the span of two years does not show any significant change after revegetation (Table 7.21).

According to Jha et al. (2001), concentration of radium was recorded in the tailing pond area was 220 Bq kg^{-1} to 1200 Bq kg^{-1} and adjacent area concentration of radium in soil up to the depth of (0–15cm) was 65 Bq kg^{-1} to 186.18 Bq kg^{-1}. Markose, 1990, Raghavayya et al. (1990), Khan et al. (2000) and Markose et al. (1981) had given similar results. In the present study after covering the uranium tailings with 30 cm of soil, the concentration of radium recorded was very low in comparison to past data given earlier.

The analysis of concentration of radium in the soil samples shows that the content of radium in the soil varied with sites and depths. It was found to be highest in quantity in experimental containers' soil, closely followed by soil of the tailing pond. Statistically, soils of experimental containers and the tailing pond are similar in radium content. As already mentioned, the quantity of radium varies significantly with depths. It has been observed that it shows an increasing trend with increasing depth. The difference between depth 0–15 and 15–30 is not significant. Statistically,

TABLE 7.21
Concentration of Radium (Bq/kg) of Soil and Tailings in Both *ex Situ* and *In Situ* Experimental Trials

Year	Site	Depth (cm)			Mean
		Soil		Tailing	
		0–15	15–30	30–45	
	Exp. containers	7.70	8.40	44.90	**20.30**
1st Year	Tailing pond	9.30*	11.90	34.20*	**18.50**
	Mean	*8.50*	*10.20*	*39.60*	*19.40*
	Exp. containers	7.20	7.70	43.40	**19.40**
2nd Year	Tailing pond	8.50	10.40	32.20**	**17.10**
	Mean	*7.90*	*9.10***	*37.80*	*18.30*
Mean of depth		**8.20**	**9.60**	**38.70**	**18.80**
Mean of Exp. Site		Exp. Containers 19.90		Tailing Pond 17.80	
	Depth	Exp. Site	Year	Dept*Exp. Site	
CD	5.44	4.44	4.44	7.70	
	Depth*Year	Exp. Site*Year	Depth*Exp. Site*Year		
	7.70	6.28	10.88		

Note: * Significance at 5% level ** Significance at 1% level *** Significance at 0.01% level •Tailings

TABLE 7.22

Concentration of Radium (Bq/kg) of Soil and Tailings in Different Experimental Sites

Site	Depth (cm)			Mean
	0–15	15–30	30–45	
Tailing pond (*in situ*)	9.32	11.90	34.25*	**18.49**
Village land	9.32**	11.90	12.32	**11.11**
Forest land	5.98	6.36	6.48	**6.27**
Exp. Containers (*ex situ*)	7.69	8.44	44.90***	**20.34**
Mean	**8.08**	**9.65**	**24.49**	**14.07**
CD	Depth	Exp. Site	Depth* Exp. Site	
	4.250	4.908	8.501	

Note: * Significance at 5% level ** Significance at 1% level *** Significance at 0.01% level •Tailings

it may be regarded as similar to radium; it was the third depth where radium was significantly higher than the first two depths, as mentioned earlier. According to Jha et al. (2007), in the Bagjata mine site, concentration of radium was recorded at 213.5 Bq kg^{-1} near the village areas.

Radium concentration in soils showed that its concentration was very similar to other radionuclides discussed earlier. It was also found that depth was an influencing factor for the concentration of radium. It was observed that the variation among three depths was highly significant. It was noted that the value of soil radium increases with the depth of the soils in both sites. Its quantity in deeper soil, i.e., 30–45 cm, was found to be nearly four times more (38.70 Bq kg^{-1}) than that of the upper soil layers (8.20 Bq kg^{-1}), i.e., up to 15 cm. It was also observed that variation in soil radium values over the span of two years does not show any significant change after revegetation.

The analysis of quantity of radium in the soil samples showed that the content of radium in the soil varied with sites and depths. It was found to be highest in quantity in experimental containers' soils, closely followed by soils of the tailing pond. As already mentioned, the quantity of radium varied significantly with depths. It has been observed that it showed an increasing trend with increasing depth. The difference between depth 0–15 cm and 15–30 cm was not significant. Statistically, it may be regarded as the same in radium; it was the third depth where radium was significantly higher than the first two depths (Table 7.22).

7.6.1 RADIONUCLIDES UPTAKE STUDY IN ROOTS AND SHOOTS OF REVEGETATED SPECIES IN BOTH *EX SITU* (EXPERIMENTAL CONTAINERS) AND *IN SITU* (TAILING POND) FOR EXPERIMENTAL TRIALS

Plant samples from the revegetated tailing pond (*in situ* trial) and from the experimental trials under *ex situ* conditions were analyzed for estimation of radionuclides.

Concentrations of radionuclides in different parts of the selected plant species were evaluated. Soil and tailing samples at different depths adhering to the root of the plants were also analyzed to estimate the concentration ratio. In certain cases when the individual for radiochemical analysis entire aboveground parts were analyzed and reported as rest parts/shoot. Analytical results of vegetation samples were reported on a dry weight basis. The order of selectivity was much more for the root, and the least accumulation was observed in the shoot.

7.6.2 Concentration of Natural Uranium (Bq/Kg) in Roots of Selected Species in Both *In Situ* and *Ex Situ* Experimental Trials in Both Ages

The analysis of uranium in roots indicates that there was a significant difference between sites, age and species. It has been found that amount of uranium is significantly (46.14%) higher in plants grown under *in situ* conditions (22.505 Bq/kg) than in *ex situ* conditions (15.40 Bq/kg) in tailing pond areas. The average value for both ages of experimentation show that the level of uranium in root increased significantly (19.3%) in the second age of experimentation (20.627) in comparison to first age's level (17.278) (Table 7.23).The amounts of uranium in root among the seven species

TABLE 7.23

Concentration of Uranium (Bq/Kg) in Roots of Selected Species in Both *Ex Situ* and *In Situ* Experimental Trials in Both Ages

Age	Sites	*Colebrookea oppositifolia*	*Dodonaea viscosa*	*Furcraea foetida*	*Imperata cylindrica*	*Jatropha gossypifolia*	*Pogostemon benghalense*	*Saccharum spontaneum*	Mean	Mean site
1st age (one year old)	Exp. Container	24.125	4.975	3.125	28.250	3.625	13.650	12.525	**12.896**	Exp. Container 15.40
	Tailing Pond	34.725	7.975	0.335	36.150	0.738	59.425	12.275	**21.660**[***]	
	Mean	*29.425*	*6.475*	*1.730*	*32.200*	*2.181*	*36.538*	*12.48*	*17.278*	
2nd Age (two year old)	Exp. Container	28.950	13.350	12.175	29.325	4.275	15.875	21.375	**17.904**	Tailing Pond 22.505
	Tailing Pond	35.975	8.625	3.475	35.625	2.125	61.250	16.375	**23.350**	
	Mean	*32.463*[*]	*10.988*	*7.825*	*32.475*[**]	*3.200*	*38.563*	*18.875*	*20.627*	
	Mean	30.944	8.731	4.777	32.338	2.691	37.550	15.637	18.953	

Critical Difference (CD) Age = 0.2265 Site = 0.2265 Species = 0.4237
Age*Site = 0.3203 Age*Species = 0.5991 Site*Species = 0.5991
Age*Site*Species = 0.8473

Note:[*] Significance at 5% level [**] Significance at 1% level [***] Significance at 0.01% level

have been found to vary significantly. It was found that the species *Pogostemon benghalense* contains the maximum amount of uranium in its roots (37.55 Bq/kg) in dry weight of root, while the species *Jatropha gossypifolia* was found to contain the least amount of uranium (2.691 Bq/kg) among the seven species. Concerning the total amount of uranium in roots, it was found that the species *Pogostemon benghalense* (37.55 Bq/kg), *Imperata cylindrica* (32.338 Bq/kg) and *Colebrookea oppositifolia* (30.944 Bq/kg) contain the maximum amount of uranium in their roots, while the species *Saccharum spontaneum* (15.637 Bq/kg), *Dodonaea viscosa* (8.731 Bq/kg), *Furcraea foetida* (4.777 Bq/kg) and *Jatropha gossypifolia* (2.691 Bq/kg) are found to contain the least amount of uranium accumulated in their roots. The combined effect of age and sites shows that uranium in plant roots in both sites increased in the second age group as compared to the first age group. During both ages, the uranium concentration in roots of selected plant species in tailing ponds remains higher than in experimental containers. Statistical analysis of the data from both the site and species interactions leads to the conclusion that all the species (in spite of variation among them) contain higher uranium when grown on tailing pond. The interaction values differ highly significantly among themselves.

The effect of the three factors on uranium accumulation in roots leads to the similar conclusion as discussed earlier. The species *Pogostemon benghalense* shows higher accumulation values of uranium in root and *Jatropha gossypifolia* shows the lowest accumulation values as compared to other species in both age groups, both under *in situ* and *ex situ* trials. All the species show an increase in uranium level after one year, i.e., in the second year, and have higher values of uranium in the tailing pond site.

7.6.3 Concentration of Natural Uranium (Bq/Kg) in Shoots of Selected Species in Both *In Situ* and *Ex Situ* Experimental Trials in Both Ages

The statistical analysis of uranium in the shoot indicates that uranium behaves in a different pattern in different species. It has been found that there exists significant difference in uranium between two sites; uranium is significantly less in experimental containers (5.836 Bq/kg) than in tailing pond (6.067 Bq/kg), i.e. a decrease of 4% (Table 7.30). The age-wise records showed that uranium has increased in the second age of experimentation. It was 4.712 Bq/kg in the first age while in the second age it reached the level of 7.191 Bq/kg. It can be observed that there has been a rise of 52.61% in uranium level in the second year. The difference was highly significant.

Also, the seven species have been found to vary significantly in respect of uranium in their shoots. It was found that species *Saccharum spontaneum* possesses the maximum amount of uranium in their shoots (9.356 Bq/kg), while the species *Jatropha gossypifolia* contains the least uranium (1.410 Bq/kg) in their shoots among the seven species. It may be observed that the species *Saccharum spontaneum* contains 6.64 times more uranium than that contained by the species *Jatropha gossypifolia*.

The interaction results of age and site show that uranium in experimental containers was very low as compared to that of tailing pond areas in the first age group, but in the second age group uranium was higher in experimental containers. The age-wise performance of the species indicates that the uranium in all the species shows an increase in the second age group. The interaction values vary significantly. A look at

TABLE 7.24

Concentration of Uranium (Bq/Kg) in Shoot of Selected Species under *Ex Situ* and *In Situ* Experimental Trials in Both Ages

Age	Sites	Colebrookea oppositifolia	Dodonaea viscosa	Furcraea foetida	Imperata cylindrica	Jatropha gossypifolia	Pogostemon benghalense	Saccharum spontaneum	Mean	Mean site
1st age (one year old)	Exp. Container	6.075	3.125	1.575	2.075	2.125	6.475	8.975	**4.346**	Exp. Container
	Tailing Pond	7.425*	2.725	0.025	11.0725	0.166	6.175	7.300	**5.077**	5.836
	Mean	*6.750**	*2.925*	*0.800*	*6.900*	*1.146*	*6.325*	*8.137*	*4.712***
2nd Age (two year old)	Exp. Container	8.125	5.625	6.625	10.625	3.125	7.525	9.625	**7.325**	Tailing Pond
	Tailing Pond	8.150	3.475	1.075	12.425	0.225	12.525	11.525	**7.057**	6.067
	Mean	*8.138*	*4.550*	*3.850*	*11.525*	*1.675*	*10.025*	*10.575****	*7.191*
	Mean	7.444	3.738	2.325	9.213	1.410	8.175	9.356	5.951
Critical Difference (CD)	Age = 0.0687 Site = 0.0687 Species = 0.1285 Age*Site = 0.0971 Age*Species = 0.1817 Site*Species = 0.1817 Age*Site*Species = 0.2570									

Note: * Significance at 5% level ** Significance at 1% level *** Significance at 0.01% level

the site-wise performance of species in respect of uranium in the shoots reveals that almost all the species contain lesser amount of uranium in experimental containers. It does well in tailing ponds. The combination values differ significantly. The overview of three-dimensional results lead to the similar conclusion already mentioned. The shoot of species *Saccharum spontaneum* contains the highest level of uranium in both sites during both years among the seven species.

7.6.4 CONCENTRATION OF NATURAL URANIUM (BQ/KG) IN SHOOTS PERCENT OF SELECTED SPECIES IN BOTH *IN SITU* AND *EX SITU* EXPERIMENTAL TRIALS IN BOTH AGES

The statistical analysis of percentage of uranium in the shoots indicates that uranium concentration varies in different species. It has been found that there exists significant difference in uranium percentage in shoots between two sites in species. Uranium percentage in shoots was significantly higher in experimental containers (46.42%) than in tailing ponds (29.27%) (Table 7.25). The age-wise data show that uranium percentage

TABLE 7.25

Concentration of Uranium Percentage in Shoots of Selected Species under *Ex Situ* and *In Situ* Experimental Trials in Both Ages

Age	Sites	*Colebrookea oppositifolia*	*Dodonaea viscosa*	*Furcraea foetida*	*Imperata cylindrica*	*Jatropha gossypifolia*	*Pogostemon benghalense*	*Saccharum spontaneum*	Mean	Mean site
1st age (one year old)	Exp. Containers	25.18	62.63	50.38	7.34	58.57	47.44	71.66	**46.20**	Exp. Container **46.42**
	Tailing Pond	21.38	34.17	7.54	32.49	22.59*	10.39	59.47	**26.86**	
	Mean	*23.28*	*48.50*	*28.96*	*19.91*	*40.58**	*28.91*	*65.56*	*36.53*	
2nd Age (two year old)	Exp. Container	28.07	42.13	54.41	36.23	73.21	47.41	45.03	**46.64**	Tailing Pond **29.87**
	Tailing Pond	22.65	40.30	30.95	34.88	10.56	20.45**	70.39	**32.88**	
	Mean	*25.36*	*41.22*	*42.68****	*35.55*	*41.88*	*33.93*	*57.71*	*39.76*	
	Mean	**24.32**	**44.86**	**35.82**	**27.73**	**41.23**	**31.42**	**61.64**	**38.15**	

Critical Difference (CD)	Age = 0.864 Site = 0.864 Species = 1.617
	Age*Site = 1.223 Age*Species = 2.287 Site*Species = 2.287
	Age*Site*Species = 3.234

Note: * Significance at 5% level ** Significance at 1% level *** Significance at 0.01% level

in shoots has increased in the second age of experimentation. It has 36.53% in the first age group while in the second age group it has reached to the level of 39.76%, and this difference is statistically significant (Table 7.25). Also, seven species have been found to vary significantly in respect of uranium percentage in their shoots. It is found that the species *Saccharum spontaneum* possesses the maximum amount of uranium percentage in their shoots, while the species *Colebrookea oppositifolia* contains the least uranium percentage in the shoots among the seven species. The interaction results of age and site show that uranium percentage in shoots in experimental containers is more than in the tailing pond area during both ages.

The age-wise performance of the species indicates that the uranium percentage in the shoots in all the species show an increase in the second age, and interaction values vary significantly. A look at the site-wise performance of species in respect of uranium in the shoots reveals that almost all the species contain a higher percentage of uranium in the shoots in experimental containers, with the exception of *Saccharum spontaneum*. It performs well on the tailing pond. The combination values differ significantly. The overview of statistical results leads to a similar conclusion. The shoots of *Saccharum spontaneum* contain the highest level of uranium in both sites during both ages among the seven species (Table 7.24).

7.6.5 CONCENTRATION OF NATURAL URANIUM (BQ/KG) IN WHOLE PLANT OF SELECTED SPECIES IN BOTH *IN SITU* AND *EX SITU* EXPERIMENTAL TRIALS IN BOTH AGES

The radioactive elements absorbed by the plants of seven different species grown in two experimental areas were estimated for two consecutive years to ascertain the status of radionuclides in whole plants as well as the percentages of soil radionuclides in whole plants. Thus, the variation in the level of radionuclides depended on experimental area, species and time. Hence, the data so recorded were analyzed as factorial analyses using statistical packageGenStat-32 (Table 7.26). The analysis reveals that the level of uranium differs significantly between the two experimental areas. It was observed that the average level of uranium in plants grown on tailing pond was significantly higher than those grown in experimental containers. Among the species,

TABLE 7.26

Concentration of Uranium (Bq/Kg) in Whole Plant of Selected Species under *Ex Situ* and *In Situ* Experimental Trials in Both Ages

Age	Sites	*Colebrookea oppositifolia*	*Dodonaea viscosa*	*Furcraea foetida*	*Imperata cylindrica*	*Jatropha gossypifolia*	*Pogostemon benghalense*	*Saccharum spontaneum*	Mean	Mean site
1st age (one year old)	Exp. Container	29.950	7.975	4.525	30.075	5.525	20.300	21.525	**17.125**[**]	Exp. Container
	Tailing Pond	42.275	10.650	0.450	46.525[*]	0.929	65.475	19.525	**26.547**	**21.118**
	Mean	*36.112*	*9.312*	*2.487*	*38.300*	*3.227*	*42.887*	*20.525*	*21.836*	
2nd Age (two year old)	Exp. Container	37.125	18.650	18.575	40.300	7.125	23.525	30.475	**25.111**	Tailing Pond
	Tailing Pond	44.125	12.075	4.525	48.075	2.225	74.175	26.675	**30.268**	**28.407**
	Mean	*40.625*	*15.362*	*11.550*	*44.188*	*4.675*	*48.850*	*28.575*[***]	*27.689*	
	Mean	38.369	12.337	7.019	41.244	3.951	45.869	24.550	24.763	
Critical Difference (CD)	Age = 0.2158 Site = 0.2158 Species = 0.4038 Age*Site = 0.3052 Age*Species = 0.5711 Site*Species = 0.5711 Age*Site*Species = 0.8076									

Note:[*] Significance at 5% level [**] Significance at 1% level [***] Significance at 0.01% level

it was found that the level of uranium varies highly significantly and *Pogostemon benghalense* (45.869 Bq/kg) contains the highest level of uranium while *Jatropha gossypifolia* (3.951 Bq/kg) contains the lowest level of uranium (Table 7.26).

The age-wise average shows that in general the level of uranium in the experimental areas has increased significantly in second age as compared to the first age. The overall average of uranium estimated during the second age was significantly greater than the level of uranium observed during the first age.

7.6.6 Concentration of Natural Uranium Percentage in Whole Plants of Selected Species in Both *In Situ* and *Ex Situ* Experimental Trials in Both Ages

The analysis showed that the percentage of soil uranium in the whole plant differs significantly between the two experimental areas. It was observed that the average percentage of uranium in plants grown in tailing pond was significantly higher than that grown in experimental containers. It was found that the percentage of uranium varied highly significantly among the species, and *Pogostemon benghalense* contains the highest percentage of soil uranium (10.864%) while *Jatropha gossypifolia* contains lowest percentage of uranium (0.956%). It may be explained that *Pogostemon benghalense* was more susceptible to absorption of uranium than other species whereas *Jatropha gossypifolia*, *Furcraea foetida*, *Dodonaea viscosa* and *Saccharum spontaneum* have poor absorption ability, or in other words it may be said that these species are least affected by the presence of a radionuclide element, while *Pogostemon benghalense* and *Colebrookea oppositifolia* are more affected or had more accumulation in their root (Table 7.27).

TABLE 7.27

Concentration of Uranium Percentage in Whole Plant of Selected Species under *Ex Situ* and *In Situ* Experimental Trials in Both Ages

Age	Sites	Colebrookea oppositifolia	Dodonaea viscosa	Furcraea foetida	Imperata cylindrica	Jatropha gossypifolia	Pogostemon benghalense	Saccharum spontaneum	Mean	Mean site
1st age (one year old)	Exp. Container	7.244	1.929	1.095	7.275	1.336	4.910	5.207	**4.142**	Exp. Container 5.131
	Tailing Pond	9.515	2.397	0.101	10.472*	0.209	14.737	4.395	**5.975**	
	Mean	*8.380*	*2.163*	*0.598*	*8.873*	*0.773*	*9.823*	*4.801*	*5.059*	

(Continued)

TABLE 7.27   (Continued)

Age	Sites	Colebrookea oppositifolia	Dodonaea viscosa	Furcraea foetida	Imperata cylindrica	Jatropha gossypifolia	Pogostemon benghalense	Saccharum spontaneum	Mean	Mean site
2nd Age (two year old)	Exp. Container	9.048	4.545	4.527	9.822	1.736	5.733	7.427	6.120***	Tailing Pond
	Tailing Pond	10.754	2.943	1.103	11.716	0.542	18.077	6.501	7.377	6.676
	Mean	*9.901**	*3.744*	*2.815*	*10.769*	*1.139***	*11.905*	*6.964*	*6.748*	
	Mean	9.140	2.954	1.706	9.821	0.956	10.864	5.882	5.903	

Critical Difference (CD) Age = 0.0526 Site = 0.0526 Species = 0.0983
Age*Site = 0.0743 Age*Species = 0.1391 Site*Species = 0.1391
Age*Site*Species = 0.1967

Note: * Significance at 5% level ** Significance at 1% level *** Significance at 0.01% level

The age-wise average showed that in general the intake of percentage uranium in the experimental areas by the species has increased significantly in the second age as compared to the first age. The overall average of percentage intake of uranium by plants during the second age was significantly higher than the percentage of uranium absorbed during the first age.

7.6.7 Concentration of Polonium (Bq/Kg) in Roots of Selected Species in Both *In Situ* and *Ex Situ* Experimental Trials in Both Ages

It was found from the statistical analysis that there was a significant difference among sites, ages and species. The data reveal that amount of polonium was significantly (82.87%) higher in tailing pond (20.381 Bq/Kg) than in experimental containers (11.14 Bq/kg). The averages of both ages of experimentation show that the level of polonium in root has increased significantly in the second age of experimentation (15.86 Bq/kg) in comparison to first age level (15.65 Bq/kg), i.e. a nominal but statistically significant increase (Table 7.28).

The amounts of polonium in roots have been found to vary significantly. It was found that *Saccharum spontaneum* contains the maximum amount of polonium in its roots (36.62 Bq/kg), while *Jatropha gossypifolia* is found to contain the least amount of polonium (4.37 Bq/kg) among the seven species. The combined effects of age and sites show that concentration of polonium in roots in both sites has increased in the second age group as compared to the first age group. During both ages, the concentration of polonium in roots of selected species in *in situ* experimental site remain higher than in *ex situ*. It conforms to the overall findings of the sites, i.e., *in situ*

TABLE 7.28

Concentration of Polonium (Bq/Kg) in Roots of Selected Species in Both *Ex Situ* and *In Situ* Experimental Trials in Both Ages

Age	Sites	Species							Mean	Mean site
		Colebrookea oppositifolia	Dodonaea viscosa	Furcraea foetida	Imperata cylindrica	Jatropha gossypifolia	Pogostemon benghalense	Saccharum spontaneum		
1st age (one year old)	Exp. Container	10.140	18.355	6.455	8.680	0.855	14.235	18.455	11.0250	Exp. Container 11.1454
	Tailing Pond	12.235	22.025	8.890	13.340	7.760	23.340	54.445	20.2907	
	Mean	**11.1875**	**20.1900**	**7.6725**	**11.0100***	**4.3075**	**18.7875**	**36.4500**	**15.6579**	
2nd Age (two year old)	Exp. Container	10.165	18.495	7.020	8.715	1.035	14.480	18.950	11.2657	Tailing Pond 20.3811
	Tailing Pond	12.345	22.145	8.925	13.425	7.855	23.950	54.655	20.4714	
	Mean	**11.255**	**20.3200**	**7.9725**	**11.0700**	**4.4450**	**19.2150****	**36.8025**	**15.8686**	
	Mean	**11.2213**	**20.2550**	**7.8225*****	**11.0400**	**4.3763**	**19.0013**	**36.6263**	**15.7632**	

Critical Difference (CD): Age = 0.01905 Site = 0.01905 Species = 0.03564
Age*Site = 0.02694 Age*Species = 0.05040 Site*Species = 0.05040
Age*Site*Species = 0.07127

Note: * Significance at 5% level ** Significance at 1% level *** Significance at 0.01% level

(tailing pond) was higher than the *ex situ* (experimental containers). The interaction values differ highly significantly among themselves. Viewing the effect of the three factors of age, site and species on polonium concentration in roots leads to the same conclusion as discussed earlier. The species *Pogostemon benghalense* shows higher values of polonium and *Furcraea foetida* the lowest values as compared to other species in both age groups in *in situ* experimental sites. All the species show an increase in polonium level after the first age, i.e., in the second age.

7.6.8 CONCENTRATION OF POLONIUM (BQ/KG) IN SHOOTS OF SELECTED SPECIES IN BOTH *IN SITU* AND *EX SITU* EXPERIMENTAL TRIALS IN BOTH AGES

The statistical analysis for the levels of polonium in the shoots of different plant species reveals that there exists significant difference between the age as well as sites, *viz.* tailing pond, experimental containers and species. The comparison of the average polonium level shows that there was a significant increase in the second age of experimentation. During the first age, the amount of polonium in shoots was found to be 9.684 Bq/kg, whereas in the second age it reached to 10.095 Bq/kg (Table 7.29).

TABLE 7.29

Concentration of Polonium (Bq/Kg) in Shoots of Selected Species in Both *Ex Situ* and *In Situ* Experimental Trials in Both Ages

Age	Sites	Colebrookea oppositifolia	Dodonaea viscosa	Furcraea foetida	Imperata cylindrica	Jatropha gossypifolia	Pogostemon benghalense	Saccharum spontaneum	Mean	Mean site
1st age (one year old)	Exp. Container	7.760	12.655	3.455	5.455	0.300	8.845	4.255	6.1036	Exp. Container
	Tailing Pond	7.760*	12.655	6.605	11.130	6.6555	12.445**	35.600	13.2643	6.4200
	Mean	*7.760*	*12.655*	*5.0300*	*8.2925*	*3.4775*	*10.6450*	*19.9275*	*9.6839*	
2nd Age (two year old)	Exp. Container	7.880	12.790	6.705	5.900	0.2900	9.045	4.545	6.7364	Tailing Pond
	Tailing Pond	7.855	12.725	6.875	11.425	6.950	12.490	35.855	13.4536	13.3589
	Mean	*7.8675*	*12.7575***	*6.7900*	*8.6625*	*3.6200*	*10.7675*	*20.200****	*10.0950*	
	Mean	7.8137	12.7062	5.9100	8.4775	3.5487	10.7062	20.0638	9.8895	
Critical Difference (CD)		Age = 0.03035 Site = 0.03035 Species = 0.05678 Age*Site = 0.04292 Age*Species = 0.08030 Site*Species = 0.08030 Age*Site*Species = 0.11356								

Note: * Significance at 5% level ** Significance at 1% level *** Significance at 0.01% level

Further, it showed that the polonium in shoots in tailing pond (13.359 Bq/kg) was significantly higher than that of experimental containers (6.420 Bq/kg). It can be observed that the level of polonium in tailing pond was more than twice that of experimental containers.

The species differ significantly among themselves in respect of polonium content in shoots. *Saccharum spontaneum* possesses the highest amount of polonium (20.0638 Bq/kg). *Jatropha gossypifolia* lags behind all the species as it was found to contain the lowest amount of polonium (3.5487 Bq/kg). From the interaction results of age and site, it may be seen that polonium in experimental containers as well as tailing pond area showed no significant change in the second age. The polonium in all the species shows a small but insignificant increase in the second age. Site-wise performance of species in respect of polonium content in the shoots showed that most of the species contain higher amount of polonium in tailing pond. The combination values differed significantly.

7.6.9 CONCENTRATION OF POLONIUM PERCENTAGE IN SHOOTS OF SELECTED SPECIES IN BOTH *IN SITU* AND *EX SITU* EXPERIMENTAL TRIALS IN BOTH AGES

The statistical analysis of percentage of polonium in the shoots reveals that there exists a significant difference in polonium percentage in shoots between the durations, each spanning two years, between two sites, *viz. in situ* (tailing pond) and *ex situ* (experimental containers). All the selected species have been found to vary significantly in respect of polonium percentage in their shoots (Table 7.30).

The comparison of the average polonium percentage in shoots showed that there was a significant increase in the second age of experimentation. During the first age, the percentage of polonium in shoots was 61.8%, whereas in the second age it reached upto 65.27%. The analysis also reveals that the percentage of polonium in shoots in tailing pond (69.47%) significantly exceeds that in experimental containers (57.61%).

As mentioned earlier, species differ significantly in respect of polonium percentage in shoots. It was found that species *Furcraea foetida* and *Imperata cylindrica* possesses *equally* the highest amount of polonium in their shoots (75.09% and 74.77%). Next in order was *Colebrookea oppositifolia* (70.28%). The species *Saccharum spontaneum* lags behind all the species in polonium percentage in shoots. It may be inferred that the species *Saccharum spontaneum* had the least uptake by polonium concentration.

From the interaction results of age and site, it may be observed that polonium percentage in shoots in experimental containers increased while in tailing pond area sit remained stable in the second age. The polonium percentage in shoots in half of the species show an increase in the second age, and the rest show neither increase nor decrease. The interaction values vary significantly. Seeing the site-wise performance of species in respect of containing polonium in the shoots, it was found that most of the species contain a higher percentage of polonium in tailing pond. The combination values differed significantly. A view on the effect of the three factors together leads to the conclusions as already discussed. The species *Saccharum spontaneum* shows

TABLE 7.30

Concentration of Polonium Percentage in Shoots of Selected Species in Both *Ex Situ* and *In Situ* Experimental Trials in Both Ages

Age	Sites	Colebrookea oppositifolia	Dodonaea viscosa	Furcraea foetida	Imperata cylindrica	Jatropha gossypifolia	Pogostemon benghalense	Saccharum spontaneum	Mean	Mean site
1st age (one year old)	Exp. Container	76.53	68.95	53.52	62.85	35.12	62.14	23.06	54.59	Exp. Container 57.61
	Tailing Pond	63.42	57.46	74.30	83.43	85.76	53.32	65.39	69.01	
	Mean	*69.98*	*63.20*	*63.91*	*73.14*	*60.44*	*57.73**	*44.22*	*61.80*	
2nd Age (two year old)	Exp. Container	77.52	69.15	95.52	67.70	28.01	62.47	23.98	60.62	Tailing Pond 69.47
	Tailing Pond	63.63	57.46	77.03	85.10	88.48	52.15	65.60	69.92	
	Mean	*70.58*	*63.31*	*86.27*	*76.40***	*58.25*	*57.31*	*44.79*	*65.27*	
	Mean	*70.28*	*63.25*	*75.09*	*74.77*	*59.34*	*57.52****	*44.51*	*63.54*	

Critical Difference (CD)

Age = 0.924 Site = 0.924 Species = 1.728
Age*Site = 1.306 Age*Species = 2.444 Site*Species = 2.444
Age*Site*Species = 3.457

Note: * Significance at 5% level ** Significance at 1% level *** Significance at 0.01% level

higher values of polonium and *Jatropha gossypifolia* the lowest values as compared to other species in both ages. The values in tailing pond are higher.

7.6.10 Concentration of Polonium (Bq/Kg) in Whole Plant of Selected Species in Both *In Situ* and *Ex Situ* Experimental Trials in Both Ages

The analysis indicated that the level of polonium in plants differs significantly between the two experimental areas. It was observed from the general averages that the average level of polonium in plants growing in tailing pond was significantly higher than the plants growing in experimental containers. Among the species, it was found that the level of polonium varied highly significantly and *Saccharum spontaneum* contains highest level of polonium (56.64 Bq/kg) while *Jatropha gossypifolia* (7.886 Bq/kg) contains the lowest level of polonium. It indicated that *Saccharum spontaneum* was more susceptible to polonium absorption (Table 7.31). The age-wise average showed that in general levels of absorption of polonium in the experimental areas by the species have increased significantly in the second age as compared to the first age. The overall average of polonium intake recorded during the second age of experimentation was significantly greater than the level of polonium observed during the first age.

7.6.11 Polonium Percentage in Whole Plant of Selected Species in Both *In Situ* and *Ex Situ* Experimental Trials in Both Ages

The analysis showed that percentage polonium in the whole plant differs significantly between the two experimental areas. It was observed that the average percentage of polonium in plants grown in tailing pond was significantly higher than that grown in experimental containers (Table 7.32).

It was found that percentage of polonium varied highly significantly among the species, and *Saccharum spontaneum* contains the highest percentage (35.524%) of polonium while *Jatropha gossypifolia* (4.998%) contains the lowest percentage of polonium. It may be explained that *Saccharum spontaneum* was more susceptible to absorption of polonium than other species, whereas *Jatropha gossypifolia* has poor absorption ability or in other words it has the least uptake by presence of polonium element while polonium uptake was maximum in *Saccharum spontaneum*. The age-wise average showed that in general the percentage intake of polonium in the experimental areas by the species has increased significantly in the second age as compared to the first age. The overall average of percentage intake of polonium by plants during second age was significantly higher than the percentage of polonium absorbed during the first age.

7.6.12 Concentration of Radium (Bq/Kg) in Roots of Selected Species in Both *In Situ* and *Ex Situ* Experimental Trials in Both Ages

The analysis of radium found in roots reveals that there is significant variation between sites, ages and species. It was observed that the amount of radium was significantly higher in *in situ* (tailing pond) experimental sites (1.07 Bq/kg) than in *ex*

TABLE 7.31

Concentration of Polonium (Bq/Kg) in Whole Plant of Selected Species in Both *Ex Situ* and *In Situ* Experimental Trials in Both Ages

Age	Sites	Species							Mean	Mean site
		Colebrookea oppositifolia	*Dodonaea viscosa*	*Furcraea foetida*	*Imperata cylindrica*	*Jatropha gossypifolia*	*Pogostemon benghalense*	*Saccharum spontaneum*		
1st age (one year old)	Exp. Container	17.950	31.010	9.960	14.110	1.130	23.030	22.735	17.132	Exp. Container 17.517
	Tailing Pond	19.790	34.605	15.545	24.345	14.480	35.670	89.995	33.490	
	Mean	*18.870*	*32.807**	*12.753*	*19.227*	*7.805**	*29.350*	*56.365*	*25.311*	
2nd Age (two year old)	Exp. Container	18.020	31.160	13.575	14.465	1.275	23.475	23.345	17.902	Tailing Pond 33.680
	Tailing Pond	20.350	34.870	15.625	24.680	14.665	36.415	90.485	33.870	
	Mean	*19.185*	*33.015*	*14.600*	*19.575*	*7.970*	*29.945*	*56.915*	*25.886*	
	Mean	**19.027**	**32.911****	**13.676**	**19.400*****	**7.887**	**29.647**	**56.640**	**25.599**	

Critical Difference (CD)   Age = 0.0541 Site = 0.0541 Species = 0.1013
Age*Site = 0.0766 Age*Species = 0.1433 Site*Species = 0.1433
Age*Site*Species = 0.2026

Note: * Significance at 5% level ** Significance at 1% level *** Significance at 0.01% level

TABLE 7.32

Concentration of Polonium Percentage in Whole Plant of Selected Species in Both *Ex Situ* and *In Situ* Experimental Trials in Both Ages

Age	Sites		Colebrookea oppositifolia	Dodonaea viscosa	Furcraea foetida	Imperata cylindrica	Jatropha gossypifolia	Pogostemon benghalense	Saccharum spontaneum	Mean	Mean site
1st age (one year old)	Exp. Container		10.5063	18.1504	5.8297	8.2587	0.7062	13.6528	13.5282	10.0276	Exp. Container 10.2856
	Tailing Pond		12.5811	21.9994	9.8824	15.4768	9.2053	22.6764	57.2123	21.2905	
	Mean		*11.5437*	*20.0749**	*7.8560*	*11.8678*	*4.9334*	*18.0780*	*35.2597*	*15.6591*	
2nd Age (two year old)	Exp. Container		10.6131	18.3521	7.9952	8.5193	0.7509	13.8259	13.7493	10.5437	Tailing Pond 21.4684
	Tailing Pond		13.0057	22.2854	9.9859	15.7730	9.3724	23.2728	57.8290	21.6463	
	Mean		*11.8094*	*20.3188*	*8.9906*	*12.1462***	*5.0617*	*18.5494*	*35.7892*	*16.0950*	
	Mean		*11.6765*	*20.1968*	*8.4233*	*12.0070*	*4.9975*	*18.3137****	*35.5244*	*15.8770*	

Critical Difference (CD) Age = 0.03418 Site = 0.03418 Species = 0.06394
Age*Site = 0.04833 Age*Species = 0.09042 Site*Species = 0.09042
Age*Site*Species = 0.12788

Note: * Significance at 5% level ** Significance at 1% level *** Significance at 0.01% level

situ (experimental containers) (0.82 Bq/kg), i.e., 30.68% higher. The age-wise aver-ages show that the level of radium in root has increased significantly in the second age of experimentation (0.97 Bq/kg) in comparison to the first age level (0.92 Bq/kg), i.e., an increase of 5.69% (Table 7.33).

The amounts of radium in root among the seven species have been found to vary significantly. It is found that the species *Furcraea foetida* possesses the least amount of radium in its roots (0.15 Bq/kg), while the species *Imperata cylindrica* contains the maximum radium concentration in roots (1.02 Bq/kg) among the seven species. The joint effects of ages and sites show that radium in both sites have increased in the second age as compared to the first age. During both ages, radium in roots of selected species in *in situ* remains higher than *ex situ*. It conforms to the overall findings of the sites, i.e., *in situ*>*ex situ*.

From the site-wise performance of species in respect of radium in the roots, it can be inferred that all the species contain a higher percentage of radium in tailing pond. The interaction values differ highly significantly. An overall view on the effect of the three factors leads to the same conclusion as discussed above. The species *Furcraea foetida* shows higher values of radium and *Imperata cylin-drica* the lowest values as compared to other species in both ages in both sites. All the species show a rise in radium level after one age group i.e., in second age group. The species possess higher values of radium in *in situ* (tailing pond) experimental site.

7.6.13 Concentration of Radium (Bq/Kg) in Shoots of Selected Species in Both In Situ and Ex Situ Experimental Trials in Both Ages

The analysis of radium in the shoots reveals that there exists significant difference in radium between two sites. Radium was significantly more in tailing pond (0.225 Bq/kg) than in experimental containers (0.138 Bq/kg) i.e. 63% more in tailing pond as compared to experimental containers (Table 7.34).

The averages of the ages showed that the amount of radium decreased in the sec-ond age of experimentation as compared to the first age, although the difference is statistically nonsignificant. It was 0.193 Bq/kg in the first age while in the second age it reduced to the level of 0.170 Bq/kg. The interaction values recorded vary signifi-cantly. This is in contrast to polonium and uranium, where a significant increase was recorded in the second age.

The amounts of radium in shoots of different species have been found to vary significantly. It was found that the species *Saccharum spontaneum* contains the max-imum amount of radium in its shoots (0.388 Bq/kg), while the *Jatropha gossypifolia* contains the least amount of radium (0.045 Bq/kg) among all the tried species. It was seen that *Saccharum spontaneum* holds 8.62 times more radium as compared to the *Jatropha gossypifolia*. However, results indicate that all the species have a higher amount of radium in the tailing pond as compared to that of experimental containers. The values varied highly significantly. From the prior analyses of uptake and accu-mulation of radionuclides in the seven species, it was seen that *Jatropha gossypifo-lia* and *Furcraea foetida* have the lowest uptake and accumulation of radionuclides,

TABLE 7.33

Concentration of Radium (Bq/Kg) in Root of Selected Species in Both *Ex Situ* and *In Situ* Experimental Trials in Both Ages

Age	Sites	Species							Mean	Mean site
		Colebrookea oppositifolia	*Dodonaea viscosa*	*Furcraea foetida*	*Imperata cylindrica*	*Jatropha gossypifolia*	*Pogostemon benghalense*	*Saccharum spontaneum*		
1st age (one year old)	Exp. Container	1.235	0.8450	0.1150	0.955	0.1000	1.135	1.145	0.7900	Exp. Container 0.8229
	Tailing pond	1.555	1.425	0.195	1.045	0.1350	1.155	1.880	1.0557	
	Mean	*1.395*	*1.135*	*0.155*	*1.000*	*0.1175*	*1.145*	*1.5125*	*0.9229*	
2nd Age (two year old)	Exp. Container	1.235	1.135	0.0950	0.990	0.1350	1.145	1.255	0.8557	Tailing Pond 1.0754
	Tailing pond	1.655	1.445	0.2250	1.1150	0.1450	1.1900	1.8900	1.0950	
	Mean	*1.445*	*1.2900**	*0.1600*	*1.0525*	*0.1400*	*1.1675*	*1.5725*	*0.9754*	
	Mean	**1.4200**	**1.2125**	**0.1575**	**1.0262**	**0.1287****	**1.1562**	**1.5425**	**0.9491*****	

Critical Difference (CD) Age = 0.01168 Site = 0.01168 Species = 0.02186

Age*Site = 0.01652 Age*Species = 0.03091 Site*Species = 0.03091

Age*Site*Species = 0.04371

Note: * Significance at 5% level ** Significance at 1% level *** Significance at 0.01% level

TABLE 7.34

Concentration of Radium (Bq/Kg) in Shoots of Selected Species in Both *Ex Situ* and *In Situ* Experimental Trials in Both Ages

Age	Sites	Colebrookea oppositifolia	Dodonaea viscosa	Furcraea foetida	Imperata cylindrica	Jatropha gossypifolia	Pogostemon benghalense	Saccharum spontaneum	Mean	Mean Site	
1st age (one year old)	Exp. Container	0.105	0.135	0.025	0.135	0.045	0.145	0.355	0.135	Exp. Container	0.138
	Tailing Pond	0.135	0.680	0.090	0.155	0.045	0.255	0.390	0.250		0.138
	Mean	*0.120*	*0.407**	*0.057*	*0.145***	*0.045*	*0.200*	*0.373*	*0.193*		
2nd Age (two year old)	Exp. Container	0.110	0.145	0.025	0.145	0.045	0.155	0.355	0.140	Tailing Pond	0.225
	Tailing Pond	0.145	0.190	0.095	0.190	0.045	0.290	0.450	0.201		0.225
	Mean	*0.128*	*0.168*	*0.060*	*0.168*	*0.045*	*0.223*	*0.403*	*0.170*		
	Mean	*0.124*	*0.287***	*0.059*	*0.156*	*0.045*	*0.211*	*0.388****	*0.181*		

Critical Difference (CD)   Age = 0.0767 Site = 0.0767 Species = 0.1434
Age*Site = 0.1084 Age*Species = 0.2029 Site*Species = 0.2029
Age*Site*Species = 0.2869

Note: * Significance at 5% level ** Significance at 1% level *** Significance at 0.01% level

while *Saccharum spontaneum* and *Pogostemon benghalense* have highest accumulation of radionuclide substances.

7.6.14 CONCENTRATION OF RADIUM PERCENTAGE IN SHOOTS OF SELECTED SPECIES IN BOTH *IN SITU* AND *EX SITU* EXPERIMENTAL TRIALS IN BOTH AGES

The analysis of radium percentage in the shoots reveals that there exists significant difference in radium percentage in shoots of selected species between the two sites. Radium percentage was significantly more in tailing pond (25.2%) than in experimental containers (20.5%). The age-wise records showed that radium percentage decreased significantly in the second age of experimentation as compared to the first age. It was 24.5% in the first age while in the second age it reduced to the level of 21.3%. The difference was statistically significant. This was quite different from the cases of polonium and uranium, where an increase was recorded in the second age (Table 7.35).

Radium percentages in shoots among the seven species have been found to vary significantly. It was found that the species *Furcraea foetida* along with *Jatropha gossypifolia* possess the maximum amount of radium percentage in their shoots (34.3% and 35.6%), while the species *Imperata cylindrica* contains the least radium (15.2%) among the seven species. It may be recalled that species *Imperata cylindrica* contains the maximum percentage of polonium (74.77%).

The joint effects of years and sites showed that radium percentage in both the sites have fallen in the second age as compared to first age. The year wise records of the species in respect of radium percentage indicated that the radium percentage in all species decreases in the second age group of plant shoots. The interaction values vary significantly. The site-wise performance of species in respect of radium in the shoots reveals that almost all the species contain a higher percentage of radium in their shoots in tailing pond. The combination values differ highly significantly. An overall view on the effects of the three factors depicts the same picture as discussed earlier. The species *Jatropha gossypifolia* shows higher values of radium percentage and *Imperata cylindrica* the lowest values as compared to other species. All the species show a fall in radium level after one year, i.e., in the second age group. The species possess higher values of radium in tailing pond.

7.6.15 CONCENTRATION OF RADIUM (BQ/KG) IN WHOLE PLANT OF SELECTED SPECIES IN BOTH *IN SITU* AND *EX SITU* EXPERIMENTAL TRIALS IN BOTH AGES

The analysis indicates that the level of radium in plants differs significantly between the two experimental sites. It was observed from the general averages that the average level of radium in plants growing in tailing pond was significantly higher than the plants growing in experimental containers (Table 7.36). From the analysis, it was also noted that among the species, it was found that the level of radium varied highly significantly and *Saccharum spontaneum* contains highest level of radium (1.924 Bq/kg) while *Jatropha gossypifolia* contains lowest level of radium (0.160 Bq/kg). It indicated that *Saccharum spontaneum* is more susceptible to radium absorption. The

TABLE 7.35

Concentration of Radium Percentage in Shoots of Selected Species in Both *Ex Situ* and *In Situ* Experimental Trials in Both Ages

Age	Sites	Colebrookea oppositifolia	Dodonaea viscosa	Furcraea foetida	Imperata cylindrica	Jatropha gossypifolia	Pogostemon benghalense	Saccharum spontaneum	Mean	Mean site	
1st age (one year old)	Exp. Container	8.5	16.0	22.5	14.1	44.9	12.8	31.0	21.4	20.5	Exp. Container
	Tailing Pond	8.7	47.1	46.0	14.8	33.3	22.1	20.7	27.5		
	Mean	*8.6*	*31.5**	*34.3*	*14.5***	*39.1*	*17.4*	*25.9***	*24.5*		
2nd Age (two year old)	Exp. Container	8.9	12.8	26.1	14.6	33.3	13.5	28.3	19.7	25.2	Tailing Pond
	Tailing Pond	8.8	13.1	42.5	17.0	31.0	24.4	23.8	22.9		
	Mean	*8.8*	*13.0*	*34.3*	*15.8*	*32.1*	*19.0*	*26.0*	*21.3*		
	Mean	*8.7*	*22.2*	*34.3*	*15.2*	*35.6*	*18.2*	*26.0****	*22.9*		

Critical Difference (CD)   Age = 5.32 Site = 5.32 Species = 9.96
Age*Site = 7.53 Age*Species = 14.09 Site*Species = 14.09
Age*Site*Species = 19.92

Note: * Significance at 5% level ** Significance at 1% level *** Significance at 0.01% level

TABLE 7.36

Concentration of Radium (Bq/Kg) in Whole Plant of Selected Species in Both *Ex Situ* and *In Situ* Experimental Trials in Both Ages

Age	Sites	Colebrookea oppositifolia	Dodonaea viscosa	Furcraea foetida	Imperata cylindrica	Jatropha gossypifolia	Pogostemon benghalense	Saccharum spontaneum	Mean	Mean site
1st age (one year old)	Exp. Container	1.335	0.677	0.105	1.090	0.140	1.255	1.525	0.875	Exp. Container 0.932
	Tailing Pond	1.665	1.555	0.255	1.225	0.155	1.435	2.245	1.219	
	Mean	*1.500**	*1.116*	*0.180*	*1.157*	*0.147*	*1.345**	*1.885*	*1.047*	
2nd Age (two year old)	Exp. Container	1.335	1.255	0.130	1.135	0.155	1.275	1.635	0.989	Tailing Pond 1.245
	Tailing Pond	1.775	1.635	0.295	1.255	0.190	1.455	2.290	1.271	
	Mean	*1.555*	*1.445*	*0.212*	*1.195*	*0.173*	*1.165*	*1.962*	*1.130*	
	Mean	1.527	1.281**	0.196	1.176	0.160	1.355	1.924***	1.088	

Critical Difference (CD)   Age = 0.0426 Site = 0.0426 Species = 0.0426

Age*Site = 0.0603 Age*Species = 0.1128 Site*Species = 0.1128

Age*Site*Species = 0.1596

Note: * Significance at 5% level ** Significance at 1% level *** Significance at 0.01% level

yearly average showed that in general levels of absorption of radium in the experimental areas by the species have increased significantly in the second age group as compared to the first age group.

The overall average of radium intake recorded during the second age of experimentation was significantly greater than the level of radium observed during the first age.

7.6.16 RADIUM PERCENTAGE IN WHOLE PLANT OF SELECTED SPECIES IN BOTH *IN SITU* AND *EX SITU* EXPERIMENTAL TRIALS IN BOTH AGES

The analysis showed that the percentage of radium in the whole plant differs significantly between the two experimental areas. It was observed that the average percentage of radium in plants grown in tailing pond was significantly higher than that grown in experimental containers (Table 7.37).

It was found that the percentage of radium varied highly significantly among the species, and *Saccharum spontaneum* contains the highest percentage of soil radium (10.334%) while *Jatropha gossypifolia* contains the lowest percentage of radium (0.856%). It may be explained that *Saccharum spontaneum* was more susceptible to absorption of radium than the other species and had the maximum radionuclide uptake, whereas *Jatropha gossypifolia* has poor absorption ability, or in other words has the least uptake by presence of radionuclide element.

Uptake and accumulation of radionuclides in the seven species observed showed that *Jatropha gossypifolia* and *Furcraea foetida* have the lowest uptake and accumulation of radionuclides, while *Saccharum spontaneum* and *Pogostemon benghalense* have the highest accumulation of radionuclide substances. Works carried out by several researchers lead one to infer that different parts of the same plant show marked differences in concentrations of elements, which probably change with time. This has been observed for nutrient elements as well as nonessential elements. In most of the literature reviewed, it was found that radium, in general, has been reported to decrease within various plants as roots > stem and leaves > seed/fruits. D'Souza and Mistry (1971) reported that the distribution pattern of radium was similar to that of calcium within different tissues of the same plant.

The perusal of data (Table 7.35) shows that radium percentages in shoots among the seven species have been found to vary significantly.

According to Jha et al., (2001), the highest concentration among all the species analyzed was found in the different organs of the *Ipomoea aquatica* (family Convolvulaceae). Concentration in root, stem and branch and leaf was 4044, 464 and 442 Bqkg^{-1}. Concentration in stem and branch and leaf was comparable to the root of other species like *Typha latifolia* and *Saccharum spontaneum* growing over a tailing pile. The plant may thus be taken as a hyperaccumulator of ^{226}Ra. In the present study the yearly average showed that in general levels of absorption of radium in the experimental areas by the species have increased significantly in the second age as compared to the first age. The overall average of radium intake recorded during the second age of experimentation was significantly greater than the level of radium observed during the first age.

Percentage of radium varied highly significantly among the species, and *Saccharum spontaneum* contains the highest percentage of soil radium (10.334%) while

TABLE 7.37
Radium percentage in *Ex Situ* and *In Situ* Experimental Trials in Both Ages

Age	Sites		Colebrookea oppositifolia	Dodonaea viscosa	Furcraea foetida	Imperata cylindrica	Jatropha gossypifolia	Pogostemon benghalense	Saccharum spontaneum	Mean	Mean site
							Species				
1st age (one year old)	Exp. Container		6.563	3.331	0.516	5.359	0.688	6.170	7.497	4.303	Exp. Container
	Tailing Pond		9.005	8.410	1.379	6.625	0.838	7.761	12.142	6.595	4.653
	Mean		*7.784*	*5.870*	*0.948**	*5.992*	*0.763*	*6.965**	*9.820*	*5.449*	
2nd Age (two year old)	Exp. Container		6.757	6.352	0.658	5.744	0.784	6.453	8.275	5.003	Tailing Pond
	Tailing Pond		10.404	9.584	1.729	7.356	1.114	8.529	13.423	7.449	7.022
	Mean		*8.580*	*7.968*	*1.194***	*6.550*	*0.949*	*7.491****	*10.849***	*6.226*	
Mean			8.182	6.919	1.071	6.271	0.856	7.228	10.334	5.837	
Critical Difference (CD)			Age = 0.2107 Site = 0.2107 Species = 0.3941 Age*Site = 0.2979 Age*Species = 0.5574 Site*Species = 0.5574 Age*Site*Species = 0.7883								

Note: * Significance at 5% level ** Significance at 1% level *** Significance at 0.01% level

TABLE 7.38

Concentration of Radium (Bq/Kg) in Whole plant of selected species in both *ex situ* and *in situ* experimental trials in both ages.

Age	Sites	Colebrookea oppositifolia	Dodonaea viscosa	Furcraea foetida	Imperata cylindrica	Jatropha gossypifolia	Pogostemon benghalense	Saccharum spontaneum	Mean	Mean Site
1st age (one year old)	Exp. Container	1.335	0.677	0.105	1.090	0.140	1.255	1.525	0.875	Exp. Container 0.932
	Tailing Pond	1.665	1.555	0.255	1.225	0.155	1.435	2.245	1.219	
	Mean	*1.500*	*1.116****	*0.180*	*1.157*	*0.147*	*1.345*	*1.885*	*1.047*	
2nd Age (two year old)	Exp. Container	1.335	1.255	0.130	1.135	0.155	1.275	1.635	0.989	Tailing Pond 1.245*
	Tailing Pond	1.775	1.635	0.295	1.255	0.190	1.455	2.290	1.271	
	Mean	*1.555*	*1.445*	*0.212*	*1.195*	*0.173******	*1.165*	*1.962*	*1.130*	
	Mean	*1.527*	*1.281*	*0.196*	*1.176*	*0.160*	*1.355*	*1.924*	*1.088*	

Critical Difference (CD) Age = 0.0426 Site = 0.0426 Species = 0.0798
Age*Site = 0.0603 Age*Species = 0.1128 Site*Species = 0.1128
Age*Site*Species = 0.1596

Note: * Significance at 5% level ** Significance at 1% level *** Significance at 0.01% level

Jatropha gossypifolia contains the lowest percentage of radium (0.856%). It may be explained that *Saccharum spontaneum was* more prone to absorption of radium than other species and had the maximum radionuclide uptake, whereas *Jatropha gossypifolia* has poor absorption ability, or in other words had the least uptake of radionuclides in comparison to other selected species.

To conclude, only the following 13 species have been screened on the basis of growth study. Further, the ethnobotanical study in and around Jaduguda shows that except for *Cleome viscosa*, locals do not use any of the species for their day-to-day requirements. To avoid translocation of radionuclides through the food chain, a trial of *Cleome viscosa* is not recommended since seeds of *Cleome viscosa* are used for culinary purposes. Therefore, only 12 species, viz., *Alysicarpus monilifer, Argemone mexicana, Cenchrus ciliaris, Cenchrus setigerus, Desmodium gangeticum, Dicanthium annulatum, Heteropogon contortus, Indigofera trita, Iseilema laxum, Ocimum canum, Rhynchosia minima* and *Tephrosia purpurea*, are recommended for revegetating tailing ponds.

8 Summary

Environmental degradation is a by-product of all mining operations, and uranium mining and milling are no exception. In uranium mining and milling operations, uranium tailings are produced as solid and liquid wastes. Millions of tonnes of these wastes are created each year due to the low grade of the majority of the world's uranium reserves. The uranium tailings are frequently disposed of locally, employing locations with favorable characteristics to build tailing basins. Uranium tailings could pose environmental risks if the disposal site is left unattended after the uranium mill has been shut down. In the form of radon emission, windblown dust distribution, and the leaching of pollutants, including heavy metals into the water, these tailings pose major dangers to the environment and human health. The main goal of an ecological strategy for radioactive waste consolidation is to manage and treat all solid and liquid waste in such a way that no living or nonliving element of the environment is harmed by any nuclear traces. Due to the possibility of radionuclides and trace metals entering the food chain, ecologically appropriate vegetation or plants that exhibit low radionuclide and trace element uptake are most suited for radiation consolidation over short periods of time over the tailings. Because of their limited root system penetration and ability to endure severe environments, they become desirable plantation candidates for remedied and covered tailings (Venu Babu, 2007).

Uranium naturally occurs, with an average concentration of 2.8 parts per million in the topmost layer of the earth. Despite the fact that it can be found in practically all rocks and soils, the World Nuclear Association (2021) estimates that there are currently 5.5 million metric tonnes of identified uranium resources available globally, with an additional 10.5 million metric tonnes likely still undiscovered. More than 4.5 billion tonnes of uranium are found in seawater, but its extraction is not yet viable economically. Depending on technological advances for recovery, 35 million tonnes are categorized as natural resources with logical possibilities for eventual economic extraction, but it is predicted that just 5.5 million tonnes of uranium exist in ore deposits that are commercially viable for extraction. However, in comparison to other metals, concentration of uranium is very low.

The environmental and public health effects linked with uranium mining have made it a contentious practice. The traditional methods of mining rock ores, that is, surface mining, heap leaching and under ground mining, aiming for maximum output to meet the rising demand have resulted in significant environmental, financial and social problems. According to Srivastava et al. (2020), the treatment of mine tailings, which contain thorium, radium, uranium and polonium and release radon, is a big problem. The unavoidable water-intensive extraction procedure is also a serious concern.

Currently, 46% of uranium is obtained through conventional mines (open pit and underground), 50% is obtained by *in situ* leaching and 4% is obtained as a by-product

DOI: 10.1201/9781003326151-8

from the extraction of other minerals. Mining techniques are always evolving. Up until 1990, underground mines produced 55% of the world's output; by 1999, this had drastically decreased to 33%; however, by 2000, new Canadian mines had raised it once more. Both normal tailings (i.e., leached ore rock leftovers) and solids produced on-site by waste treatment procedures can be considered solid wastes at U mill tailings facilities. Sludge or "mud" from neutralizing acidic mine/mill effluents, which contains Fe and a variety of coprecipitated elements, or barium sulphate precipitates that specifically remove Ra are examples of the latter (Campbell et al., 2015). Fifty-five percent of the world's uranium production was done through *in situ* leaching, compared to 38% from open pit and underground mines and 7% from production as a by-product. Due mostly to Kazakhstan, *in situ* leach (also known as *in situ* recovery, or ISR) mining has been continuously growing its proportion of the total, and in 2020 it accounted for over half of production.

The inherent radioactivity of uranium and the resulting pollution with radioactive dust, radon gas, water-borne poisons and elevated background radiation levels create more severe effects than do other forms of mining. Higher environmental contamination results from these radioactive wastes produced during mining, whether they take the form of mill tailings or mining waste. The majority of uranium mines worldwide typically yield poor quality ores with 0.1% to 0.3% U_3O_8. India's mines have still lower grades than these. As a result, the uranium business produces a lot of trash. After recovering the minor amounts of uranium it contained, the mined ore almost all ends up as waste, contaminating wide areas with radioactive waste. The inherent radioactivity of uranium has therefore more severe effects than other metals.

After mining is finished, tailings are permanently covered with enough rock to withstand erosion and enough clay and soil to reduce gamma radiation levels and radon emission rates to levels close to those naturally occurring in the area. Then, a vegetative cover is created. Since environmental restoration involves multiple disciplines, it is necessary to establish an integrated strategy to maximize resources. Additionally, it should be mentioned that the general environmental damage in the area includes radioactive contamination (IAEA, 1997).

Wastes produced during mining and milling operations need to be restored in a way that is ecologically safe so that the plant species utilized to do so don't endanger the health of nearby residents or their livestock. It is necessary to conduct appropriate screening of plant species during restoration to ensure that neither people nor cattle consume the plants, even if only a small amount of uptake does occur. By doing this, the mobility of radionuclides in the food chain will be monitored. Furthermore, this would benefit the local community, the mining industry and the ecosystem of the site. Eco rejuvenation of these places based on an ecosystem approach is preferable to simple revegetation or bioreclamation in order to support the sustainable growth of the mined areas (Wali, 1987; Soni et al., 1994).

A variety of technical tools can be employed to stop or minimize the severity of these processes. Capping can be used to regulate radon emissions, moisture ingress and chemical processes that might promote leaching. The physical and chemical properties of the tailings can be improved *in situ* or by reprocessing to boost long-term stability. By making containment structures better, the bare minimum safety can be attained. Given that, all of these engineering solutions have a finite life span;

ecological solutions might provide sustainable ones. Plants can be used to economically resource radioactivity in soil tailings or to consolidate pollution.

Tailings and wastewater from uranium mines have substantially different physical and chemical characteristics from typical soils and streams. Because of this, many of the conventional methods for collecting and analyzing natural soils and streams do not directly apply to tailings. Second, there is a wide range in the physical and chemical properties of uranium tailings. The kind of tailing deposition, milling method, mineral composition and effluent treatment are all quite site specific and vary greatly. The residual solid slurry or tailings from the uranium ore processing industry are dumped into a specially designed system known as a tailing pond after the economically viable portion of the ore has been recovered. The tailing pond contains the highest quantitative concentration of residual uranium among radionuclides, and a number of environmental factors, including precipitation, pH changes, redox potential, microbial activity and organic associations have the ability to fix (precipitate) or solubilize it. The current investigation included the chemical fractionation of "U" in mill tailings from tailing ponds at the Jaduguda uranium mining and ore processing plant, both active and inactive. Additionally, research has been done on the impact of several physicochemical parameters (such as pH, EC, TC, etc.) on uranium mobility (Singh et al., 2018).

Potential chemical and radioactive pollution comes from the tailing solids and tailing porewater (NUTO, 1985). For vegetation to thrive and for the soil to be used sustainably, the physical characteristics of the soil are crucial, while soil fertility, plant growth and reproduction depend on its chemical composition. Because of this, it's critical to comprehend how soil chemistry and soil structure interact to affect the soil's ability to store and release nutrients. Because the depth of the rooting zone and the air and water relationships inside it are largely governed by the physical makeup of the soil, the physical characteristics of soil have a direct relationship with plant growth and establishment. Additionally, because many chemical and biological characteristics of soil fertility can be deduced from physical features, they have extra indirect significance. The important thing to remember is that all physical factors have an impact on the management techniques needed to sustain moisture potential, a range of porosity, temperature, and mechanical resistance ideal for mine lands' high production and soil growth (Munshower, 1993; Soni et al., 1992).

The mobility of weathered uranium in the soil profile depends on the soil's attraction for uranium and on soil characteristics that include its acidity, alkalinity and water content. Water should therefore be minimal in areas where soil strongly binds uranium (usually in soils high in organic matter), as these areas have a high affinity for doing so. Accordingly, mobility is probably greater in soils that bind uranium less firmly, such as semiarid climatic soils with neutral to alkaline soil mixed with low organic carbon content (Ebbs et al., 1998). Uranium content in the soil has a major impact on plant uranium concentration.

Through numerous food chains, radionuclides that are eventually mobilized within plant bodies make their way into the human system. Clay and organic matter are active components for supplying plant nutrition, according to soil science literature. Because they have negative surface charges, they have a high cationic adsorptive capacity. The pH, moisture content, cation-exchange capacity and competing ions are some of the factors that can influence the migration of radionuclides and trace elements from soil/substrate (NCRP, 1992). In addition, environmental factors like temperature and soil texture play a significant role in determining how much bioaccumulation occurs in plants. Before becoming a part of the metabolic system, radionuclides and trace elements travel over a complicated pathway.

Due to the possibility of radionuclides and trace metals entering the food chain, vegetation or plants that exhibit significant uptake of radionuclides and trace elements may be suitable for phytoremediation over short periods of time but may not be suitable for a vegetative cover directly over the tailings. However, because of their shallow root penetration and ability to endure hard conditions, they make desirable planter candidates for the soil-covered, remedied tailings. This also applies to vetiver (khas grass) (Venu Babu, 2007).

Two objective criteria of the sustainable development of land resources against the backdrop of ecological civilization are the management of land consolidation and ecological risk assessment. The subject of monitoring and managing ecological threats is receiving more and more attention. In order to identify, assess and control ecological risks based on full-process analyses of land consolidation projects, including ecosystem service functions as an imperative, it is necessary to analyze ecological risk factors and evaluation index systems relative to the ecological effect of land consolidation. This analysis must be done on the basis of systematic elaboration of the connotation and characteristics of ecological risks of the consolidation.

As a result, the constraint of earlier research on ecological risk has been overcome with a greater focus on changes to an ecosystem's structure and functioning. In order to address the ecological dangers of land consolidation, scientific challenges and development trends that require further attention are identified (Liu et al., 2017).

For the cleanup of uranium mill tailings, ecologically viable species must be chosen, and this choice must be based on the standards that require the fewest inputs and the least amount of care. The chosen plant should have deep, wide roots and, ideally, be hardy, grow quickly and sucker (Sastry and Kavathekar, 1990).

With fewer competitors than other species and disturbances allowing the germination and establishment of non seeded plants, the native vegetation may be more beneficial and effective in restoration and bioengineering programs (Munshower, 1993). Native vegetation plays a key role in both preventing landslides and stabilizing existing ones. However, plant species with shallow root systems but strong binding ability must be favored for the rehabilitation of mill tailings.

Although absorption through leaves also happens, the majority of plants absorb nutrients and pollutants from the soil solution primarily through the roots. The amount of soluble soil uranium that is available for uptake by plants depends on how tightly and strongly uranium is linked to soil components. Therefore, it is expected that the factors affecting uranium mobility in soil will have a significant impact on how contaminated plants are. Plants do absorb uranium, however at a lower rate than mobile radioactive elements such as cesium and strontium, but higher than that of plutonium and americium. The plants appear to easily absorb the soluble forms of uranium. Natural uranium can be irregularly distributed and has a low solubility in many soils. In general, the concentration of uranium in plants decreases from the root through the shoots to the fruit and seeds. Some grasses exposed to elevated levels of uranium in soil at broad concentrations have shown stimulatory effects on growth. For many matrices, including processed waste, mine waste and tailings, concentration ratios have been established (Royal Society, 2000). The concentration ratio for crops grown on various types of soil and soil uranium has been shown to vary by four orders of magnitude. This considerable range substantially limits the use of generic models that take plant uptake into account and emphasizes the need-based approach with clearly defined source terms and composition (Rodriguez-Freire et al., 2021).

In a review of the literature from uranium botanical prospecting in the Colorado plateau, it was noted that when either sulphur or selenium is added, there is an increase in uranium's solubility in the carnotite plot. Uranium uptake was thought to be in the anionic complex form (Cannon, 1960). This was thought to be the result of oxidation in the soil solution to uranyl sulphate and uranyl selenate. Seven species that are good soil binders and acceptable species for tailing pond areas were the focus of Singh and Soni's (2010a) research. These species were all effective soil binders. Vasistha (1992) studied the root penetration and soil binding capacity of primary colonizing species on rock phosphate mines of Doon (India) and concluded that *Buddleja asiatica*, *Wendlandia exserta* and *Trema politoria* are effective soil binders and are suitable soil-conserving species on degraded mined lands.

Initially harsh physical characteristics prevailing on the study site have been favorably and conspicuously modified both by planting selective species for uranium tailing sites and by the time following plant introduction. Indorante et al., (1981) also found that after reconstruction, the properties of newly constructed mine soils reflected premining overburden character and the method of soil construction.

Many plants change the soil and the soil biota, which may change how the soil forms and how plants thrive. There are two main pathways that the connection between plants and soil operates on, according to Wardle et al., (2004): (i) associated with roots and (ii) associated with litter. Interactions with various root symbionts and additional rhizosphere biota make up the portion of the plant-soil feedback system related to roots and root exudates (Bonkowski et al., 2000). Plants may experience both positive and negative effects from this biota (Hinsinger et al., 2003; Jones and Smith, 2004). The beneficial organisms mostly consist of fungal symbionts and symbiotic nitrogen fixing bacteria (Smith and Read, 2008). To measure the bioproductivity of a natural ecosystem or a restored site, the component of plant growth that is important is the net primary production or total yield. For growth analysis, two measurements are required: (1) the plant weight, which is typically the oven dry weight (kg), but which may also refer to the organic matter or energy content; and (2) the size of the assimilatory system, which is typically the leaf area (m2) but which may also refer to the leaf protein or chlorophyll content.

A model system for examining biotic interactions underground and aboveground is provided by the ecological restoration of mine tailings through revegetation. In a tailings ecosystem undergoing restoration, microbes have been demonstrated to react quickly to environmental change. In tight correlation with changes in the physicochemical and biological conditions during the ecological restoration of mine tailings, microorganisms may be potential bioindicators for tracking the functions of the soil ecosystem.

At sites with hazardous, radioactive waste, the mobility of radionuclides and their subsequent off-site dispersion are crucial elements for risk assessments and selecting corrective measures. The uptake and deposit of plant matter to surface soil is one potential route for transfer from the soil column. A key factor for plant uptake is the behavior and mobility of radionuclides in soil, which is regulated by a variety of factors (Romney et al., 1974; Poston et al., 1995a). Radionuclide uptake characteristics have been seen to change based on the soil's texture, organic matter content, bacterial activity, pH, redox potential and physiochemical speciation, among other factors (Poston et al., 1995b; White and Dunaway, 1975; Romney et al., 1974; Han and Lee, 1997; Salbu and Oughton, 1995).

The four main categories of modern remediation technologies are physical, chemical, biological and thermal. Contaminants will be eliminated by thermal treatment, but other soil properties that are good for plant growth will be lost. The quality of the soil can also be diminished by chemical extraction and soil washing. The ability of bioremediated soils to support the development of grass and trees after being amended with green waste compost may offer a sustainable alternative for the restoration of contaminated soils or degraded land. A natural process called phytoremediation is carried out by plants that can survive in contaminated media.

Plants known as hyperaccumulators are able to concentrate high concentrations of pollutants within their roots, branches and leaves. The removal of heavy metals from soil frequently involves a variety of plant species. An inexpensive *in situ* method for restoration, revival and reuse is the use of soil amendments. Singh and Soni (2010b) claim that tree species with aboveground height and sustaining deep roots are responsible for radon emission from tailings and are therefore not suitable for consolidation. Instead, Singh and Soni advocate plants with shallow root systems and limited crown coverage.

Given the waste fixation process, the two most important radionuclides that might be found in the environmental matrices are U (nat) and ^{226}Ra. The latter is a bone seeker with behavior comparable to calcium, whereas the former is chemically more toxic and accumulates in the kidney (a known essential element for bone development). Consequently, these two elements are also the most important radionuclides in the uranium processing sector in terms of abundance, known chemical/radiological toxicity and extended half-lives. Some uranium emissions take place during the processing and extraction of uranium ore. The radium and other uranium decay products released at mines and ore processing facilities are typically more significant than the uranium in tailing ponds (Wikipedia, 2022).

The Uranium Corporation of India (UCIL), which is owned by the Indian Atomic Energy Department, runs the uranium mines in India. These are situated in the Indian tribal region of Jharkhand, in the villages of Jaduguda, Bhatin and Narwarpahar. The cost estimates for the various stabilization techniques were included in the case study on tailing ponds along with field-testing of these various methods. The consolidation of radioactivity in tailings must be done in a way that is ecologically safe so that the plant species utilized not only cover the tailings with greenery but are also environmentally sustainable on the one hand and do not endanger the health of nearby humans and livestock on the other. The case study that was presented sought to:

- assess the ecological impact of uranium tailings on soil and vegetation,
- evaluate the radionuclide uptake by vegetation growing on reclaimed tailing sites,

- evaluate the distribution of radionuclides in tailing ponds at different depths, and
- identify the physicochemical characteristics of soil and tailings.

The study was undertaken to develop a protocol for sustainable stabilization of radioactivity using native species of grass, herbaceous and shrub species. The main objectives the study were:

- selection and propagation of species of nonedible and nondomestic nature with good conservation value and
- phytostabilization of uranium mill tailings to prevent their atmospheric dispersal and adverse impact on the environment.

Additionally, diverse species' development patterns make efficient use of the soil's moisture while giving the soil caps the best vegetation cover throughout the year. To create a plant community for the aim of phytostabilization, it is necessary to adhere to specific cultural practices:

1. The plant species picked should be able to tolerate the radiological and physicochemical conditions of the mill tailings.
2. The amount of the metal or trace element in the plants above ground should not exceed the toxicity thresholds for domestic animals (NRC, 2005b).
3. The concentration of the metal/radionuclide should not be higher than the leaf/tissue toxicity limits.
4. For the plant community to be better established, appropriate sowing or transplantation must be done.
5. Until the plants are established, periodic irrigation and the addition of some manure or minimal fertilisation are required (Munshower, 1994; Tordoff et al., 2000; Williams and Currey, 2002).
6. It is necessary to engage in a certain level of cultural and agronomic practises, utilising either herbicides or physical removal of undesirable species, particularly tree species. Otherwise, such unmanaged mill wastes may eventually get colonised by invasive plant species.
7. The plant community needs to include plant species that have a greater leaf area index and can tolerate drought and field fires.

This study was conducted to determine how native primary colonizing plant species could concentrate the radioactivity in the tailings. The tailing pond was covered with a 30 cm layer of dirt as a result of this program. For the purpose of studying radionuclide uptake, seeds of the chosen plant species were dispersed over the soil-covered tailing pond (*in situ* trials). Additionally, an effort was made to produce the same species *ex situ* in 200-litertest containers. Both above- and belowground plant samples, soil samples, and tailing samples were gathered at various depths at study locations referred to as tailing pond (*in situ* experiment) and experimental containers (*ex situ* experiment) located at the Health Physics Unit campus in Jaduguda. Aside from the tailing pond, samples of soil were also taken from nearby village and forest land to compare the physical and chemical characteristics of the soils. After analyzing the samples, the mean values over the previous two years were calculated. After increasing vegetative cover, natural levels of uranium, polonium and radium were detected in all of the locations' soil as well as in a variety of selected plant species. The main challenges in replanting these soils with tailings are the concentrations of uranium, polonium, radium and other radionuclides. However, planting or sowing native species that have been acclimated typically produces favorable environmental circumstances for plant growth. Although there was a case study in Jaduguda tailing sites with very low radioactive levels, it was done to prevent any long-term effects from these tailings. The tailings have been covered with a layer of soil that is 30 cm thick in order to prevent any long-term effects of these tailings on the environment, humans, cattle, and native flora and wildlife. Gamma radiation and radon emission levels decreased as a result. However, the region has been replanted using plant species with shallow root systems, high conservation value and low canopy cover in order to consolidate the radioactivity in the tailings on a sustainable basis. The research site's initially unfavorable physical characteristics have been noticeably and favorably changed both through the introduction of certain plant species for uranium tailing sites and over the course of time. To prevent radionuclide diffusion in any form in the environment, radionuclide uptake by the chosen plants for covering the tailings must be minimal. Other essential selection criteria include the need that grown plants have no ethnobotanical significance in the nearby villages.

References

Adejuwon, J. O. and Ekanade, O. (1988). Soil changes consequent upon the replacement of tropical rain forest by plantations of Gmelina arborea, Tectona grandis and Terminalia superba. *Journal of World Forest Resource Management*, 3: 47–59.

Adewole, M. B. and Adesina, M. A. (2011). Impact of marble mining on soil properties in a part of Guinea Savanna zone of Southwestern Nigeria. *Ethiopian Journal of Environmental Studies and Management*, 4(2): 1–8.

Anon. (1982). *Report on a Workshop on the Measurement of Soils-to-Plant Transfer Factors for Radionuclides, Part 1.* Wageningen: International Union of Radioecologists, Prepared by Euratomital.

Anonymous (2015–16). Uranium Production from Mines during the Last Decade (in Tonnes) world nuclear association. *The Nuclear Fuel Report 2015*, 2017 London, United Kingdom.

Attribution-Share Alike License. Uranium Mining: Creative Commons Attribution. https://creativecommons.org/

Aweto, A. O. (1981). Secondary succession and soil fertility restoration in South Western Nigeria. I. Succession. *Journal of Ecology*, 69: 601–607.

Babu, V. (2007). *Study of Vetiver for Uptake of Uranium from Tailings.* Unpublished Personal Communication.

Bassegio, C., Campagnolo, M. A., Schwantes, D., Gonçalves Junior, A. C., Manfrin, J., Schiller, A. D. P. and Bassegio, D. (2020). Growth and accumulation of Pb by roots and shoots of *Brassica juncea* L. *International Journal of Phytoremediation*, 22: 134–139.

Bastias, J. G. (1987). *Retreatment of Radioactive Gold Bearing Tailings and Rehabilitation of Mill and Tailings Dump Sites at Rockhole and Moline, Northern Territory (a Personal View).* In Australian Mining Industry Council Environment Workshop Papers (Adelaide, September 21–25, 1987), Australian Mining Industry Council, Canberra. Proceedings of Conference "Reclamation, A Global Perspective", Calgary, Alberta, August 27–31, 1989, pp. 319–346.

Basu, S. K., Jha, V. N. and Khan, A. H. (2000). *Uptake of Radionuclides and Trace Metals by Plants Growing on or Near Uranium Tailings.* Ninth National Symposium on Environment, Bangalore University, India, pp. 138–141, June 5–7.

Baumann, N., Arnold, T. and Haferburg, G. (2014). Uranium contents in plants and mushrooms grown on a uranium-contaminated site near Ronneburg in Eastern Thuringia/ Germany. *Environmental Science and Pollution Research*, 21: 6921–6929. https://doi.org/10.1007/s11356-013-1913-5

Bauer, A., Berg, W. A. and Gold, W. C. (1978). Correction of nutrient deficiencies and toxicities in strip mined lands in semi arid and arid regions. In Schalter, F. A. (ed) *Reclamation of Drastically Disturbed Lands*. Madison, pp. 451–456, WI: American Society of Agronomy.

Beadle, C. L. (1985). Plant growth analysis. In *Techniques in Bioproductivity and Photosynthesis.* 2nd ed. Pergamon International Library of Science, Technology, Engineering and Social Studies.

Belcher, J. W., Keddy, P. A. and Twolan-Strutt, L. (1995). Root and shoot competition intensity along a soil depth gradient. *Journal of Ecology*, 83: 673–682.

Bhimaya, C. P. and Seri, D. N. (1980). Root system and root ecology. In Seri, D. N. (ed) *Environment and Root Behaviour. Geobios International.* Jodhpur, pp. 1–24.

Blume, H. P. and Röper, H. P. (1977). Der Mineralbestand als bodengenetischer Indikator. *Mittlgn. Dtsch. Bodenkundl. Gesellsch.*, 25: 797–824.

Bohm, W. (1979). *Method of Studying Root Systems*. Berlin: Springer-Verlag.

Boileau, L. J. R., Beckett, P. J., Lavoie, P., Richardson, D. H. S. and Nieboer, E. (1982). Lichens and mosses as monitors of industrial activity associated with uranium mining in Northern Ontario, Canada—part 1: Field procedures, chemical analysis and inter-species comparisons. *Environmental Pollution Series B, Chemical and Physical*, 4(1): 69–84.

Bollard, E. G. (1960). Transport in the xylem. *Annual Review of Plant Physiology*, 11: 141–166.

Bonkowski, M., Cheng, W., Griffiths, B. S., Alphei, J. and Scheu, S. (2000). Microbial-faunal interactions in the rhizosphere and effects on plant growth. *European Journal of Soil Biology*, 36(3–4): 135–147.

Bradshaw, A. D. (1987). The reclamation of derelict land and the ecology of ecosystems. In Jordan III, W. R., Gilip, M. E. and Aber, J. D. (ed) *Restoration Ecology*. Cambridge, England: Cambridge University Press, pp. 53–74.

Bradshaw, A. D., Dancer, S. W., Handley, J. F. and Sheldon, J. C. (1975). The biology of land revegetation and reclamation of the China clay wastes of Cornwall. In Chadwick, M. J. and Goodman, G. T. (ed) *The Ecology of Resource Degradation and Renewal*. Symposium of the British Ecological Society 15. Oxford: Blackwell Scientific Publications, pp. 363–384.

Bradshaw, A. D., Humphries, R. N., Johnson, M. S. and Roberts, R. D. (1975). The restoration of vegetation on derelict land produced by industrial activity. In Holdgate, M. W. and Woodman, M. J. (eds) *The Breakdown and Restoration of Ecosystems*. New York: Plenum Press, pp. 249–274.

Brugge, D., deLemos, J. L. and Oldmixon, B. (2005). Exposure pathways and health effects associated with chemical and radiological toxicity of natural uranium: A review. *Reviews on Environmental Health*, 20(3): 177–194.

Brynes, W. R. and Miller, J. H. (1973). Natural revegetation and cast overburden properties of surface mined coal lands in Southern Indiana. In Hutnik, R. J. and Devis, G. (eds) *Ecology and Reclamation of Devastated Land*. Vol. 1. New York: Gordon and Breach, pp. 285–305.

Bussler, B. H., Brynes, W. R., Pope, P. E. and Chaney, W. R. (1979). Properties of mine soil reclaimed for forest land use. *Soil Science Society American Journal*, 48: 178–184.

Callaway, J. C., Borde, A. B. and Diefenderfer, H. L. (2012). Pacific Coast tidal wetlands. In Batzer, D. P. and Baldwin, E. H. (eds) *Wetland Habitats of North America: Ecology and Conservation Concerns*. Berkeley, CA: University of California Press.

Campbell, K. M., Gallegos, T. J., & Landa, E. R. (2015). Biogeochemical aspects of uranium mineralization, mining, milling, and remediation. Applied Geochemistry, 57, 206–235.

Cannon, H. L. (1952). The effect of uranium vanadium deposits on the vegetation of the Colorado Plateau. *American Journal of Science*, 250: 735–770.

Cannon, H. L. (1960). The development of botanical methods of prospecting for uranium on the Colorado plateau. *Geological Survey Bulletin*, 1085A: 19–20.

Carter, L. J. (2015). *Nuclear Imperatives and Public Trust: Dealing with Radioactive Waste*. Routledge RFF Press, Taylor & Francis group.

Carvalho, F. P., Oliveira, J. M. and Malta, M. (2014). Radioactivity in soils and vegetables from uranium mining. *Procedia Earth and Planetary Science*, 8: 38–42.

Carvalho, F. P., Oliveira, J. M., Malta, M. and Eugenia Lemos, M. (2014). Radioanalytical assessment of environmental contamination around non-remediated uranium mining legacy site and radium mobility. *Journal of Radioanalytical and Nuclear Chemistry*, 299(1): 119–125.

Cerne, M., Smodis, B., Strok, M. and Jacimovic, R. (2018). Plant accumulation of natural radionuclides as affected by substrate contaminated with uranium-mill tailings. *Water, Air, & Soil Pollution*, 229(11): 371.

Champion, H. G. and Seth, S. K. (1968). *A Revised Forest Types of India*. Delhi: Manager of Publications, Government of India.

Chapman, D. F. and Lemaire, G. (1993). Morphogenic and structural determinants of plant regrowth after defoliation. In Baker, M. J. (ed) *Grasslands for Our World*. Wellington: SIR, pp. 55–64.

Charro, E. and Moyano, A. (2017). Soil and vegetation influence in plants natural radionuclides uptake at a uranium mining site. *Radiation Physics and Chemistry*, 141: 200–206.

Chicchester, F. W. and Hauser, V. L. (1991). Changes in chemical properties of constructed mine soils developing under forage grass management. *Soil Science Society of American Journal*, 55: 451–459.

Ciolkosz, E. J., Cronce, R. C., Cunninggham, R. L. and Petersen, G. W. (1985). Characteristics, genesis and classification of Pennsylvania mine soils. *Soil Science*, 139: 232–238.

Clegg, J. W. and Foley, D. D. (1958). *Uranium Ore Processing Addison Wesley*. Readings, MA: Natural Background Radiation in the United States [Atoms for Peace Geneva Series].

Clewell, A. F. (1999). Restoration of riverine forest at Hall Branch on phosphate mined land, Florida. *Restoration Ecology*, 7: 1–14.

Cornwell, S. M. and Stone, E. L. (1968). Availability of nitrogen to plants in acid coal mine spoils. *Nature*, 217: 768–769.

Cornwell, W. K., Cornelissen, J. H. and Amatangelo, K. (2008). Plant species traits are the predominant control on litter decomposition rates within biomes worldwide. *Ecology Letters*, 11(10): 1065–1071.

Cortois, R., Veen, G. F. and Duyts, H. (2017). Possible mechanisms underlying abundance and diversity responses of nematode communities to plant diversity. *Ecosphere*, 8(5): e01719.

Courtney, R. (2012). Mine tailings composition in a historic site: Implications for ecological restoration. *Environmental Geochemistry and Health*, 35: 79–88.

Dadhwal, K.S. 1999. Rehabilitation of limestone mine spoils with reference to agroforestry. *Indian Journal of Agroforestry*, 1: 141–148.

Datar, A. and Patrick, A. (2011). *Post-Mined Land Rehabilitation in India: Cataloguing Plant Species Used in Land Revegetation Conference Paper*. Conference: Australian Mine Rehabilitation. Australian Mine Rehabilitation Workshop 2011, Adelaide, SA.

Davy, D. R., Conway, N. F. and Newton, P. J. F. (1974). *Australia Atomic Energy Commission Report- In the Alligato Rivers Area Fact – Finding Study*.

Dean, K. C., Froisland, L. J. and Shirts, M. B. (1986). *Utilization and Stabilization of Mineral Wastes*. US Baureau of Mines Bulletin 688. Proceedings of Conference "Reclamation, A Global Perspective", Calgary, Alberta.

Diehl, P. (2009). Uranium mining and milling wastes: An introduction. In *WISE Uranium Project*. Arnsdorf, Germany.

Diehl, P. (2011). Uranium mining and milling wastes: An introduction. In *WISW Uranium Project*. pp. 21, 25

Dietz, A. C. and Schnoor, J. L. (2001). Advances in phytoremediation. *Environmental Health Perspectives*, 109(Suppl 1): 163–168.

DOE (US Department of Energy). (1986a). *Recommendation by the Secretary of Energy of Candidate Sites for Site Characterization for the First Radioactive-Waste Repository (DOE/S-0048)*. Washington, DC: US Department of Energy.

DOE (US Department of Energy). (1986b). *A Multiattribute Utility Analysis of Sites Nominated for Characterization for the First Radioactive-Waste Repository—A Decision-Aiding Methodology* (DOE/RW-0074). Washington, DC: Office of Civilian Radioactive Waste Management, US Department of Energy.

DOE (US Department of Energy). (1986c). *Nuclear Waste Policy Act (Section 112), Environmental Assessment, Deaf Smith County Site, Texas*. US Dept.of Energy. Office of Civilian Radioactive waste managment. Washington DC 20585.

DOE (US Department of Energy). (1986d). *Draft Area Recommendation Report for the Crystalline Repository Project* (DOE/CH-15). Washington, DC: Crystalline Repository Project Office, Office of Civilian Radioactive Waste Management, US Department of Energy. Washington, DC 20585

DOE (US Department of Energy). (1986e). *Environmental Assessment Remedial Action at the Tuba City Uranium Mill Tailings Site, Tuba City, Arizona (DOE/EA-0317).* US Dept. of Energy. Office of Civilian Radioactive waste managment. Washington DC 20585.

D'Souza, T. J. and Mistry, K. B. (1970). *Comparative Uptake of $ sup 230$ Th, $ sup 226$ Ra, $ sup 210$ Pb, and $ sup 210$ Po by Plants.* Bombay: Bhabha Atomic Research Center.

D'Souza, T. J. and Mistry, K. B. (1971). *Studies on the Uptake, Distribution and Metabolic Fate of Radium in Plants.* New Delhi, India: International Symposium on Use of Isotopes and Radiation in Agriculture and Animal Husbandry Research, pp. 428–438.

Eapen, S. and D'Souza, S. F. (2005). Prospects of genetic engineering of plants for phytoremediation of toxic metals. *Biotechnology Advances*, 23: 97–114.

Eatough, J. P. and Henshaw, D. L. (1993). Radon and monocytic leukaemia in England. *Journal of Epidemiology anid Commnzuniti Health*, 47: 506–507.

Ebbs, S. D., Brady Danielle, J. and Leon Kochian, V. (1998). Role of uranium speciation in the uptake and translocation of uranium by plant. *Journal of Experimental Botany*, 49(324): 1183–1190.

EEA. (2008). *Exposure Pathways and Monitoring "Air and Health".* European environment agency.

Ehlken, S. and Kirchner, G. (2002). Environmental processes affecting plant root uptake of radioactive trace elements and variability of transfer factor data: A review. *Journal of Environmental Radioactivity*, 58: 97–112.

Engelbrecht, R. (2020). Environmental radioactivity monitoring. In *Handbook of Radioactivity Analysis: Volume 2* (pp. 1–40). Academic Press.

Eisenbud, M. (1987). *Environmental Radioactivity.* 3rd ed. San Diego, CA: Academic Press Inc.

EPA. (1994). *Technical Report Design and Evaluation of Tailings Dams (EPA 530-R-94-038 NTIS PB94-201845).* Washington, DC: US Environmental Protection Agency Office of Solid Waste.

Fagan, W. F. and Bishop, J. G. (2000). Trophic interactions during primary succession: herbivores slow a plant reinvasion at Mount St. Helens. *The American Naturalist*, 155(2): 238–251.

Fashola, M. O., Ngole-Jeme, V. M. and Babalola, O. O. (2016). Heavy metal pollution from gold mines: Environmental effects and bacterial strategies for resistance. *International Journal of Environmental Research and Public Health*, 13: 1047. https://doi.org/10.3390/ijerph13111047.

Favas, P. J., Pratas, J., Mitra, S., Sarkar, S. K. and Venkatachalam, P. (2016). Biogeochemistry of uranium in the soil- plant and water-plant systems in an old uranium mine. *Science of the Total Environment*, 568: 350–368.

Feasby, D. G. (1997). Environmental Restoration of Uranium Mines in Canada: Progress Over 52 Years in IAEA-TECDOC-982. In Proceedings of a Workshop Held Under the Planning for Environmental Restoration of Uranium Mining and Milling Sites in Central and Eastern Europe.

Fetcher, Ned, Bruce L. Haines, Roberto A. Cordero, D. Jean Lodge, Lawrence R. Walker, Denny S. Fernandez, and William T. Lawrence. "Responses of Tropical Plants to Nutrients and Light on a Landslide in Puerto Rico." *Journal of Ecology* 84, no. 3 (1996): 331–41. https://doi.org/10.2307/2261196.

Finch, W. I., Butler, A. P., Armstrong, F. C. and Weissenborn, A. E. (1972). *Nuclear Fuels: Uranium: USGS Professional Paper 820.* Published by USGS (U.S. Geological Survey). Washington, DC, pp. 456–467.

Frouz, J. (2018). Effects of soil macro-and mesofauna on litter decomposition and soil organic matter stabilization. *Geoderma*, 332: 161–172.

Fuller, W. H. (1974). Desert soils. *Desert biology: special topics on the physical and biological aspects of arid regions*, 2: 31–102.

Fyles, I. H. and Mcgill, W. B. (1987). The development of nitrogen cycling in sub alpine reclaimed mine sites in South Eastern British Columbia. *Canadian Journal of Soil Science*, 67: 117–133.

Gans, I., Futhrmanos, D., Welter, E. and Wollenhaupt. (1981). Radium in Wastewater from Coalmines and Other Sources in Federal Republic of Germany. In Proceedings of the Special Symposium II, Natural Radiation Environment, Bhaba Atomic Research Centre, Bombay (INDIA), pp. 444–451.

Gavrilescu, M., Pavel, L. V. and Cretescu, I. (2009). Characterization and remediation of soils contaminated with uranium. *Journal of Hazardous Materials*, 163(2–3): 475–510.

Ghose, M. K. (2001). Management of topsoil for geo-environmental reclamation of coal mining areas. *Environmental Geology*, 40: 1405–1410.

Ghose, M. K. (2004a). *Effect of Opencast Mining on Soil Fertility*. nopr.niscair.res.in.

Ghose, M. K. (2004b). Restoration and revegetation strategies for degraded mine land for sustainable mine closure. *Land Contamination and Reclamation*, 12: 363–378.

Ghose, M. K. and Kumar, A. (2004). Mineral industries and their environmental aspects in Indian context. *Indian Journal of Engineering and Materials Sciences*, 11: 433–437.

Ghosh, R., Sinha, T. K. and Saxena, N. C. (1998). *Environmental Issues of Wastes from Thermal Power Plants*. Proceedings of NSE-7, ISM, Dhanbad, February 5–7.

Gopalakrishnan, G., Cristina, N. M. and Snyder, S. W. (2011). A novel framework to classify marginal land for sustainable biomass feedstock production. *Journal of Environmental Quality*, 40(5): 1593–1600.

Goss, M. J. (1987). The specific effects of roots on the regeneration of soil structure. In Monnier, G. and Goss, M. J. (eds) *Soil Compaction and Regeneration*. Rotterdam: A. A. Balkema, pp. 145–155.

Grammer, E. J. (1981). The uranium mill tailings radiation control act of 1978 and NRC's agreement state program. *Natural Resources Lawyer*, 13(3): 469–522.

Groenevelt, P. H., Kay, B. D. and Grant, C. D. (1984). Physical assessment of a soil with respect to rooting potential. *Geoderma*, 34(2): 101–114.

Gunther, A., Bernhard, G., Geipel, G., Reich, T., RoBherg, A. and Nitsche, H. (2003). Uraniumspeciation in plants. *Radiochemica Acta*, 91(6): 319–328.

Gupta, R. and Sarangi, A. K. (2005). *Emerging Trend of Uranium Mining: The Indian Scenario*. Conference: IAEA International symposium on "Uranium production and raw materials for the nuclear fuel cycle-supply and demand, economics, the environment and energy security", Viena.

Han, B. S. and Lee, K. J. (1997). The effect of bacterial generation on the transport of radionuclide in porous media. *Annals of Nuclear Energy*, 24: 721–734.

Hass, J. R., Elizabeth, H. B. and William, O. P. (1998). Bioaccumulation of metals by lichens: Uptake of aqueous uranium by *Peltigera membranacea*as a function of time and pH. *American Mineralogist*, 83: 1494–1502.

Hausenbuiller, R. L. (1981). *Soil Science, Principles and Practices*. Dubuque, IA: William C. Brown.

Hauser, V. L. and Chichester, F. W. (1989). Water relation of clay mine and constructed soil profiles. *Soil Science Society of American Journal*, 1: 1189–1196.

Healey, S. P. and Gara, R. I. (2002). *The Effect of a Teak (Tectona grandis) Plantation on the Establishment of Native Species in an Abandoned Pasture in Costa Rica*.

Henshaw, D. L. (1991). Radon. *Structural Survey*, Publisher: MCB UP Ltd, 9(2): 145–151. https://doi.org/10.1108/EUM0000000003247.

Henshaw, D. L., Eatough, J. P. and Richardson, R. B. (1990). Radon as a causative factor in induction of myeloid leukaemia and other cancers, *The Lancet*, 335(8696): 1008–1012.

Heyraud, M. and Cherry, R. D. (1979). Polonium-210 and lead-210 in marine food chains. *Marine Biology*, 52: 227–236.

Hinsinger, P., Plassard, C., Tang, C. and Jaillard, B. (2003). Origins of root-mediated pH changes in the rhizosphere and their responses to environmental constraints: A review. *Plant and Soil*, 248(1–2): 43–59.

Ho, D. M. L. (2015). Study on the Applicability of Structural and Morphological Parameters on Selected Uranium Compounds for Nuclear Forensic Purposes (doctoral dissertation). www.world-nuclear.org/information-library/nuclear-fuel-cycle/mining-of-uranium/world-uranium-mining-production.aspx.

Hossner, L. R. and Hons, F. M. (1992). Reclamation of mine tailings. In Lal, R. and Stewart, B. A. (eds) *Advances in Soil Science*. Vol. 17. New York: Spring-Verlag, pp. 311–350.

Hossner, L. R., Loeppert, R. H., Newton, R. J. and Szaniszlo, P. J. (1997). *Uranium Uptake by Plants from the Brassica Family and High Biomass Crop*. The US Govt. ANRCP-Egited 97-09.

Huhle, B., Heilmmeier, H. and Merkel, B. (2008). *Potential of Brassica Juncea and Helianthus Annuus in Phytoremediation for Uranium*. Proceedings of International Conference on "Uranium Mining and Hydrogeology," Bergakademie Frieberg, Germany, UMH-V, September.

Humbel, F. X. (1975). A study of soil macro porosity bases on permeability data, application of a filtration modal for fenalitic soils of Cameroon. *Pedologie*, 13: 83–117.

IAEA. (1981). *Current Practices and Options for Confinement of Uranium Mill Tailings* (STI/DOC/10/209 ISBN 92-0-125281-1). Vienna: IAEA.

IAEA. (1987). Safe management of wastes from the mining and milling of uranium and thorium ores. *International Atomic Energy Agency Safety Series*, 85: 138–142.

IAEA. (1992). *International Atomic Energy Agency, Current Practices for the Management of Uranium Mill Tailings* (Technical Report Series No. 335). Vienna: IAEA.

IAEA. (1997). *Planning for Environmental Restoration of Uranium Mining and Milling Sites in Central and Eastern Europe*. Proceedings of a workshop held under the Technical Co-operation Project RER/9/022 on Environmental Restoration in Central and Eastern Europe, Felix, Romania, IAEA-TECDOC-982, pp. 4–8.

IAEA. (2003). *Protection of the Environment from the Effects of Ionizing Radiation*. Proceedings of an International Conference, Stockholm.

IAEA. (2004a). *Environmental Contamination from Uranium Production Facilities and Their Remediation*. Proceedings of an International Workshop, Lisbon, February 11–13.

IAEA. (2004b). *The Long Term Stabilization of Uranium Mill Tailings* (IAEA-TECDOC-1403. 1011-4289). Vienna: IAEA.

IAEA. (2006). *International Atomic Energy Agency. Joint Convention on the Safety of Spent Fuel Management and on the Safety of Radioactive Waste Management*. IAEA International Law Series No. 1. Vienna: IAEA.

IAEA. (2009). *IAEA Nuclear Energy Series No. Nw-G-1.1 Policies and Strategies for Radioactive Waste Management*. Vienna: International Atomic Energy Agency.

IARC. (2022). Ijsselmeer after the chernobyl accident. In *The Radioecology of Natural and Artificial*. IARC Agents Classified by the IARC Monographs. Vol. 1–131. WHO.

Indorante, S. J., Jansen, I. J. and Boast, C. W. (1981). Surface mining and reclamation. Initial changes in soil character. *Journal of Soil and Water Conservation*, 36(6): 347–351.

Ingested, T. (1979). Mineral nutrient requirements of Pinus sylvestris and Picea abies seedling. *Physiologia Plantarum*, 45: 370–373.

Jatin Kumar, S. (2012). *Environmental Contamination*. Edited by Jatin Kumar, S. Rijeka, Croatia: InTech Janeza Trdine 9. Technical Co-operation Project RER/9/022 on

Environmental Restoration in Central and Eastern Europe, Felix, Romania, 4–8 November 1996.

Jenny, H. (1980). Prairie-forest biosequences: The soil resource. *Ecological Studies*, 37.

Jha, G., Jha, V. N., Kumar, R., Patnaik, R. L., Srivastava, V. S. and Sethy, N. K. (2001). *Environmental Surveillance Around the Uranium Complex at Jaduguda, Jharkhand*. First Environmental Survey Laboratory Professionals Meet, Kalpakkam, India, pp. I-10A–1-19.

Jha, S., Khan, A. H. and Mishra, U. C. (2001). A study of the technologically modified sources of ^{222}Rn and its environmental impact in an Indian U mineralised belt. *Journal of Environmental Radioactivity*, 53: 183–197.

Jha, V. N., Giri, S., Paul, S., Sethy, N. K., Sukla, A. K., Singh, G., Tripathi, R. M. and Puranik, V. D. (2007). *Radionuclide Uptake by Native Vegetation Growing Around Upcoming Uranium Mining Projects at Banduhurang, Jharkhand*. Proceedings of the NSE-15 on "Mitigation of Pollutants for Clean Environment", Mumbai, pp. 575–579, June 5–7.

Jha, V. N. and Jha, G. (2001). Evaluation of effluent management practices and its environmental impact around Jaduguda after thirty year of UCIL operations. *Proceeding of the International Conference on Radiation Protection Measurement and Dosimetry: Current Practices and Future Trend*, 24(1–2): 481–483.

Jha, V. N., Sethy, N. K., Sahoo, S. K., Shukla, A. K., Tripathi, R. M. and Khan, A. H. (2005a). *A Comparison of Radioactivity Level in Discharge Waste and Natural Sources in Uranium Mineralized Areas of Singhbhum, Jharkhand*. Proceedings of 27th IARP National Conference of Occupational and Environmental Radiation Protection, BARC, Mumbai, pp. 284–286.

Jha, V. N., Sethy, N. K., Shukla, A. K., Sahoo, S. K. and Khan, A. H. (2005b). Dissolved ^{222}Rn in Hydrosphere of uranium mineralized area of Singhbhum, Jharkhand. *Journal of the Association of Environmental Geochemists*, 8(1–2): 175–179.

Jha, V. N., Tripathi, R. M., Sethy, N. K. and Sahoo, S. K. (2016). Uptake of uranium by aquatic plants growing in fresh water ecosystem around uranium mill tailings pond at Jaduguda, India. *Science of the Total Environment*, 539: 175–184.

Jones, M. D. and Smith, S. E. (2004). Exploring functional definitions of mycorrhizas: Are mycorrhizas always mutualisms? *Canadian Journal of Botany*, 82(8): 1089–1109.

Jones, M. J. and Wild, A. (1975). *Soils of West African Savana, the Maintenance and Improvement of Their Fertility*. Technical Communication No. 55 of the Commonwealth Bureau of Soils. Common Wealth Agricultural Bureau, p. 246.

Joshi, M. (1995). Seasonal changes in species diversity, phytomass and net primary production in a temperate Grassland in Kumaun, Central Himalaya, India. *Journal of Tropical Forest Science*, 74(4): 521–531.

Jovanovic, L. J., Cupac, S. and Janjic, V. (2004). *Uptake and Distribution of Uranyl Nitrate in Soybean, Sunflower and Maize Plants*. ARI SERBIA Center for Pesticides and Environmental Research Centre, Zemun Society for experimental Biology, p. 1.06.

Karaca, O., Cameselle, C. & Reddy, K. R. (2018). Mine tailing disposal sites: contamination problems, remedial options and phytocaps for sustainable remediation. *Reviews in Environmental Science and Biotechnology*, 17, 205–228. https://doi.org/10.1007/s11157-017-9453-y

Kardol, P., Martijn, B. T. and Putten, V. D. W. H. (2006). Temporal variation in plant—soil feedback controls succession. *Ecology Letters*, 9(9): 1080–1088.

Karimian, N. and Razmi, K. (1990). Influence of perennial plants on chemical properties of arid calcareous soils in Iran. *Soil Science*, 150(4): 717–721.

Kechavarzi, C. and Lord, R. A. (2009). *Expanding the Potential Biomass Crop Production: Reusing Brownfield Sites and Biodegradable Wastes*. http://www.bay-publishing.com/ article - 335

Keskin, T. and Makineci, E. (2009). Some soil properties on coal mine spoils reclaimed with black locust (Robinia pceudoacacia L.) and umbrella pine (Pinus pinea L.) in Agacli-Istanbul. *Environmental Monitoring and Assessment*, 159(1): 407–414.

Khan, A. H. (2008). *Remediation of Uranium Mill Tailings: Role of Vegetation and Plants.* Homi Bhabha Centenary, DAE-BRNS National Symposium on "Landscaping for Sustainable Environment", 20–21 November. BARC, IL-12, Mumbai, pp. 93–102.

Khan, A. H., Jha, G., Srivastava, G. K., Jha, S., Jha, V. N., Srivastava, V. S. and Patnaik, R. L. (2002). *Assessment of Radiation Exposure of Uranium Mine Workers in India.* Submitted for presentation at the IAEA International Conference on "Occupational Radiation Protection", Geneva, August 26–30.

Khan, A. H., Jha, V. N., Jha, S. and Kumar, R. (2000). *Management of Radioactivity Waste and Environmental Impact of Mining and Processing of Uranium Ore.* International Symposium on Geo- Environment Reclamation, Nagpur, pp. 20–22.

Khan, A. H. and Puranik, V. D. (2011). Radon in the environment and in dwellings. In *A Uranium Mining Area in Eastern India: An Overview*. Radiation Protection Dosimetry, pp. 1–4.

Kirk-Othmer (1969). *Encyclopedia of Chemical Technology.* Edited by Mark, H. F., Mckella, J. J. and Othmer, D. F. Vol. 21. NewYork: JohnWiley, pp. 127–133.

Kirlew, P. W. and Bouldin, D. R. (1987). Chemical properties of the Rhizosphere in an acid sub soil. *Soil Science Society of American Journal*, 51: 128–132.

Konstantinou, M., Babalonas, D. (1996). Metal uptake by Caryophyllaceae species from metalliferous soils in northern Greece. *Pl Syst Evol* 203, 1–10. https://doi.org/10.1007/BF00985233

Kumar, S., Soni, P. and Vasistha, H. B. (1997). Community structure and biomass production in a restored rock phosphate mine. *Indian Forester*, 371–378.

Lal Singh (2008). Ecological impact assessment of uranium tailings with special reference to radionuclide concentration in vegetation. Ph.D. Thesis, FRI University, Dehradun (India)

Laxman, S. K., Rao, C. M. and Sudhakar, G. (2014). Evaluation of uranium mine tailing remediation by amending land soil and invading native plant species. *IOSR Journal of Environmental Science, Toxicology and Food Technology (IOSR-JESTFT)*, 8(11) Ver. II: 64–81. www.iosrjournals.org.

Li, J., Nie, M., Powell, J. R., Bissett, A. and Elise, P. (2020). Soil physico-chemical properties are critical for predicting carbon storage and nutrient availability across Australia. *Environmental Research Letters*, 15(9): 1–11, 094088.

Lim, K. H., Maene, L., Maesschalck, G. and Wan Sulaiman, W. H. (1981). *Reclamation of Tin Tailings for Agriculture in Malaysia.* Technical Bulletin Faculty of Agriculture, Wolter Kluwer: University Pertamian Malaysia, p. 61.

Linsalata, P., Robert, M. S., Helen, F., Merril, E., Eduardo, P., de Castro, M. B., Lobao, N. Y., Ivanor, S. and Marcia, C. (1989). An assessment of soil-to-plant concentration ratios for some nature analogues of the transuranic element. *Health Physics*, 56(1): 33–46. Published by Wolter Kluwer.

Liu, S. L., Yun, H. X., Zhang, Y. Q., Yin, Y. J. and Wang, J. (2017). Assessment of and controlling strategies for ecological risks of land consolidation based on ecosystem services. *Journal of Ecology and Rural Environment*, 33(3): 193–200.

Lovey, D. R., Phillips, E. J. P., Gorby, Y. A. and Landa, E. R. (1991). Microbial reduction of uranium. *Nature*, 350: 413–416.

Macaskie, L. E., Empson, R. M., Cheetham, A. K., Grey, C. P. and Skarnuliss, A. J. (1992). Uranium bioaccumulation by a citrobacter sp. as a result of enzymically mediated growth of polycrystalline $HUO_2 PO_4$. *Nature*, 257: 782–784.

Maiti, S. K., Nandhini, S. and Das, M. (2005). Accumulation of metals by naturally growing herbaceous and tree species in iron ore tailings. *International Journal of Environmental Studies*, 62(5): 593–603.

Mann, H. and Fyfe, W. S. (1995). Uranium uptake by algae: Experimental and natural environments. *Canadian Journal of Earth Science*, 22: 1899–1903.

Markose, P. M. (1990). Studies on the Environmental Behavior of Radium from Uranium Mill Tailings (PhD thesis, University of Bombay), pp. 62, 155.

Markose, P. M., Eappan, K. P., Raghavayya, M. and Patel, B. (1981). *Bioaccumulation of Radium in an Aquatic Ecosystem*. Proceedings of International Symposium on Natural Radiation Environment (special-2), Bombay, pp. 234–238.

Marple, M. I. and Potter, L. D. (1982). *Radium – 226 in Plants and Substrate at Inactive Uranium Mill Sites the Southwestern United States*. Proceedings of the Second Special Symposium on Natural Radiation Environment held at BARC, Bombay, 1981, pp. 251–257.

Martin, P. and Ryan, B. (2004). *Nature Series Radio Nuclides in Traditional Aboriginal Foods in Tropical Northern Australia. A Review*. Environmental Research Institute of the Supervising Scientist (eriss), Australia paul. Martin @ deh. Gov. au; Bruce.Pyan@deh gov. au, February 26.

Mathur, A. K. (2005). Biomining—a promising approach. *Indian Association of Nuclear Chemists and Allied Scientists Bulletin: Environmental Biotechnology*, 4(1): 10–15.

Matthew, C., van Loo, E. N., Thom, E. R., Dawson, L. A. and Care, D. A. (2001). *Understanding Shoot and Root Development Part of the Plant Sciences Commons, and the Soil*. The XIX International Grassland Congress Took Place in São Pedro, São Paulo, Brazil. Proceedings Published by Fundacao de Estudos Agrarios Luiz de Queiroz USGS 2016.

Mbagwu, J., Lal, R. and Scott, T. W. (1983). Physical properties of three soils in southern Nigeria. *Soil Science*, 136(1): 48–55.

McIntyre, T. (2003). Consideration of plant-based remediation and restoration of contaminated sites containing heavy metals—the Canadian experience. In Šašek, Václav & Glaser, John & Baveye, Philippe (eds.) *The Utilization of Bioremediation to Reduce Soil Contamination: Problems and Solutions*. Dordrecht: Springer, pp. 295–311.

Meers, E., Vandecasteele, B., Ruttens, A., Vangronsveld, J. and Tack, F. M. (2007). Potential of five willow species (Salix spp.) for phytoextraction of heavy metals. *Environmental and Experimental Botany*, 60(1): 57–68.

Mendez, M. O. and Maier, R. M. (2008). Phytostabilization of mine tailings in arid and semi-arid environments-an emerging remediation technology. *Environmental Health Perspectives*, 116(3): 278–83.

Mengel, K. and Kirkby, E. A. (1979). *Principles of Plant Nutrition, Zed. Chapter 11: Calcium*. Berne, Switzerland: International Potash Institute.

Miera, F. R., Jr. Hansou, W. C., Glandey, E. S. and Jose, P. (1980). Mobility of elevated levels of uranium in the environment. *Natural Radiation Environment III*, 1: 681–699.

Milodowski, A. E., West, J. M., Pearce, J. M., Hyslop, E. K., Basham, I. R. and Hooker, P. J. (1990). Uranium mineralized micro organisms associated with uraniferrous hydrocarbons in South West Scotland. *Nature*, 347: 465–467.

Minguzzi, C. and Vergnano, O. (1948). II contento di nichel nelli ceneri di alyssum bertlonii desv. *Atti Della Società Toscana di Scienze Naturali*, 55: 49–77.

Misdaq, M. A., Chaouqi, A., Ouguidi, J., Touti, R. and Mortassim, A. (2015a). Radon and thoron measured in petrol and gas-oil exhaust fumes by using CR-39 and LR-115 II nuclear track detectors-radiation doses to the respiratory tract of mechanic workers. *Health Physics*, 108(6): 592–596.

Misdaq, M. A., Matrane, A. and Ouguidi, J. (2015b). Radon, thoron and progeny measured in urban health centres and the resulting radiation doses to doctors, nurses and patients from the inhalation of air. *American Journal of Environmental Protection*, 4(3): 139–151.

Mistry, K. B., Gopal- Ayengar, A. R. and Bharathan, K. C. (1965). On the radioactivity of plants from the high radiation areas of the Kerala cost and adjoining regions II. *Studies on the Uptake of Alpha and Gamma E mitters Health Physics*, 11: 1459–1470.

Morishima, H., Koga, T., Kawai, H., Honda, Y. and Katsurayama, K. (1976). Experimental studies on the movement of uranium in the soil and vegetable. *Radioisotopes*, 25: 773.

Morten, L. S., Evans, C. V. and Estes, G. O. (2002). Natural uranium and thorium distribution in podozodlized soil and native blueberry. *Journal of Environment Quality*, 31(1): 155–162.

Mudd, G. M. (2000). *Remediation of Uranium Mill Tailings Wastes in Australia: A Critical Review in Contaminated Site Remediation: From Source Zones to Ecosystems*. Proceeding 2000 CSRC, Melbourne, VIC, C. D. Johnston. Vol. 2, 777, December 4–8.

Munshower, F. F. (1993). *Practical Handbook of Disturbed Land Revegetation*. London: Lewis Publishers.

Munshower, F. F. (1994). *Plant Growth on Ash Disposal Ponds (No. CONF-9406250-)*. Manhattan, KS: Great Plains- Rocky Mountain Hazardous Substance Research Center.

Nambier, E. K. S. (1981). *Ecological and Physiological Aspects of the Development of Roots: From Nursery to Forest*. Proceedings Australia Forest Nutrition Workshop, Productivity in Perpetuity, Canberra, (SIRO, Melbourne), pp. 117–129.

Nambier, E. K. S. (1983). Root development and configuration in intensively managed radiata pine plantations. *Plant and Soil*, 71: 37–47.

Nara, K. (2006). Pioneer dwarf willow may facilitate tree succession by providing late colonizers with compatible ectomycorrhizal fungi in a primary successional volcanic desert. *New Phytologist*, 171(1): 187–198.

National Research Council. (2005a). *America's Lab Report: Investigations in High School Science*. Edited by Singer, S. R., Hilton, M. L. and Schweingruber, H. A. Committee on High School Science Laboratories: Role and Vision, National Research Council. Division of Behavioral and Social Sciences and Education. Washington, DC: The National Academies Press.

National Research Council. (2005b). *Systems for State Science Assessment*. Edited by Wilson, M. R. and Bertenthal, M. W. Committee on Test Design for K-12 Science Achievement. Board on Testing and Assessment, Center for Education. Division of Behavioral and Social Sciences and Education. Washington, DC: The National Academies Press.

NCRP. (1975). *National Council on Radiation Protection and Measurements Report No. 45*. Bethesda, MD. National Council on Radiation Protection and Measurement.

NCRP. (1992). Radiological assessment: Predicting the transport Bioaccumulation and uptake by man of radionuclides released to the environment. *National Council on Radiation Protection and Measurements*, 73–77.

Niu, Z. X., Sun, L. N., Sun, T. H., Li, Y. S. and Wang, H. (2007). Evaluation of phytoextracting cadmium and lead by sunflower, ricinus, alfalfa and mustard in hydroponic culture. *Journal of Environmental Sciences*, 19(8): 961–967.

NUTO National Uranium Tailings Program. (1985). *Uranium Tailings Sampling Manual Golder*. Associates by Feenstra, S., Reades, D. W., Cherry, J. A. and Senes. Consultants by Chambers, D. B., Case, G. G. and Ibbotson, B. G. NUTP-1E CANMET Canada Centre for Mineral and Energy Technology. National Uranium Tailings Program 1E.

Nye, P. H. (1966). The effect of nutrient intensity, buffering power of a soil, absorbing power and size of root, on hairs of a root and nutrient absorption by diffusion. *Plant and Soil*, 25: 82–85.

Oades, J. M. (1984). Soil organic matter and structural stability, mechanisms and implications for management. *Plant and Soil*, 76: 319–337.

OECD-NEA and IAEA. (2020). *Uranium Resources, Production and Demand ('Red Book')*. World Nuclear Association, The Nuclear Fuel Report 2015, 2017 and 2019. Office of the Supervising Scientist, Technical Memorandum 48. Darwin, AGPS, OECD Publishing, Paris, p. 52.

Olivares, A. R., Carrillo-González, R., González-Chávez, M. D. C. A. and Hernández, R. M. S. (2013). Potential of castor bean (Ricinus communis L.) for phytoremediation of mine tailings and oil production. *Journal of Environmental Management*, 114: 316–323.

Olsen, S. R., Watanabe, F. S. and Bowman, R. A. (1983). Evaluation of fertilizer phosphate residues by plant uptake and extractable phosphorus. *Soil Science Society of American Journal*, 47: 952–958.

Pain, C. F. and Ollier, C. D. (1995). Inversion of relief-a component of landscape evolution. *Geomorphology*, 12: 151–165.

Pamela, G. and Sue, M. (2014). Using ecological attributes as criteria for the selection of plant species under three restoration scenarios. *Austral Ecology*, 39(8).

Pandey, V. C., Bajpai, O., Pandey, D. N. and Singh, N. (2015). Saccharum spontaneum: an underutilized tall grass for revegetation and restoration programs. *Genetic Resources and Crop Evolution*, 62(3): 443–450.

Pandey, V. C., Singh, K., Singh, J. S., Kumar, A., Singh, B. and Singh, R. P. (2012). Jatropha curcas: a potential biofuel plant for sustainable environmental development. *Renewable and Sustainable Energy Reviews*, 16(5): 2870–2883.

Panwar, P. and Bhardwaj, S. D. (2000). Performance of shrubs in sand and limestone mines of H.P. *Indian Forester*, 126: 1921–1925.

Parthiban, K. T. and Rai, R. S. V. (1992). Effects of a few plant species on soil physical properties. *Journal of Tropical Forest Science*, 6(3): 223–229.

Patnaik, R. L. (2002). *Assessment of Radiation Exposure of Uranium Mine Workers in India*. Submitted for Presentation at the IAEA International Conference on "Occupational Radiation Protection", August 26–30, Geneva.

Planinsek, P., Smodis, B. and Benedik, L. (2016). Simultaneous determination and uptake assessment of selected radionuclides in plants grown in substrate contaminated with U-mill tailings. *Journal of Radioanalytical and Nuclear Chemistry*, 309(1): 351–365.

Ponge, J. F., Pérès, G. and Guernion, M. (2013). The impact of agricultural practices on soil biota: a regional study. *Soil Biology and Biochemistry*, 67: 271–284.

Popova, O. N., Kodeneva, R. P. and Vavilov, P. P. (1964). Distribution in plant of radium Absorbed from the soil. *Soviet Plant Physiology*, 11: 371–375.

Poston, T. M., Antonio, E. J. and Cooper, A. T. (1995a). *Radionuclide Concentrations in Terrestrial Vegetation and Soil on and Around the Hanford Site, 1983 Through 1993*. Richland, WA: Pacific Northwest Laboratory.

Poston, T. M., Antonio, E. J. and Cooper, A. T. (1995b). *Pacific Northwest Lab*. Richland, WA: Funding Organisation; Washington, DC: USDOE.

Potter, K. N., Carter, F. S. and Doll, E. C. (1988). Physical properties of constructed and undisturbed soils. *Soil Science Society of American Journal*, 52: 1435–1438.

Power, J. F., Sandoval, F. M. and Ries, R. E. (1978). Restoration of productivity to disturbed land in the Northern Great Plains. In Wright, R. A. (ed) *The Reclamations of Disturbed Arid Lands of N. Mexico*. Albuquerque, NM: University of New Mexico Press, pp. 33–49.

Prasad, M. N. V. (2007). Phytoremediation in India. In *Phytoremediation*. Humana Press., pp. 435–454.

Pulford, I. D. and Watson, C. (2003). Phytoremediation of heavy metal-contaminated land by trees—a review. *Environment International*, 29(4): 529–540.

Punshon, T., Dickinson, N. M. and Lepp, N. W. (1996). The potential of Salix clones for bioremediating metal polluted soil. In *Heavy Metals and Trees. Proceedings of a Discussion Meeting, Glasgow*. Edinburgh: Institute of Chartered Foresters, pp. 93–104.

Raghavayya, M., Iyengar, M. A. R. and Markose, P. M. (1990). Estimation of radium -226 by emanometry. *Bulletin of Indian Association for Radiation Protection*, 3(4): 11–15.

Ran, N. L., Filin, S. and Hanan, E. (2013). Plant growth parameter estimation from sparse 3D reconstruction based on highly-textured feature points. *Precision Agriculture*, 14: 586–605.

Rankama, K. and Sahama, T. G. (1950a). *Geochemistry of Major Oxides in Host Rocks in Vizianagarm Manganese Ores Belt (A.P.), India*. Chicago, IL: University of Chicago Press, p. 912.

Rankama, K. and Sahama, T. G. (1950b). *Geochemistry of Uranium*. Chicago, IL: The University of Chicago Press, p. 640.

Rawat, A., Vasistha, H. B., & Soni, P. (2012). Ecological approach to landslide risk remediation. Ecosystem Approach to Disaster Risk Reduction, 1, 95–101.

Rayno, D. R. (1983). Estimated dose to man from uranium millling via the beef/milk foodchain pathway. *Science of the Total Environment*, 31(3): 219–241.

Rayno, D. R., Momein, M. H. and Sabau, C. (1980). *Forage Uptake of Uranium Series Radionuclides in the Vicinity of the Anaconda Uranium Mill Report Number CONF - 80115 –6*. 3rd Annual Symposium on Uranium Mill Tailings Management, November 24, 1989, IT Collins, Co USA.

Revelle, R. and Munk, W. (1977). The carbon dioxide cycle and the biosphere. In *Energy and Climate. The National Academy of Sciences*. Washington, DC, pp. 140–158.

Rickard, W. H. and Keough, R. F. (1968). Soil plant relationships of two steppe desert shrubs. *Plant and Soil*, 29: 205–215.

Rickard, W. H., Klepper, E. L., Schreckhise, R. G., Brauer, F. P., Cline, J. F., Emery, R. M., Fitzner, R. E., Paine, D., Rogers, L. E., Soldat, J. K. and Uresk, D. W. (1977). *Ecology of Waste Management Areas: Radioecology of Uranium*. B. E. Vaughn and Staff Members, Pacific Northwest Laboratory Annual Report 1976 to ERDA Assistant Administrator for Environment and Safety. Battelle, Pacific Northwest Laboratory, Richland, WA, BM-2100, pp. 4.28–4.41.

Riley, R. G., Zachara, J. M. and Wobber, F. J. (1992). *Chemical Contaminants on DOE Lands and Selection of Contaminants Mixture for Subsurface Science Research*. Process Metallurgy, 6: 991. DOE Office of Energy Report DOE/ER-0547T.

Ritcey, G. M. (1989). *Tailings Management: Problems and Solutions in the Mining Industry*. Amsterdam: Elsevier Science Publishers, B.U.

Roberts, J. A., Daniels, W. L., Bell, J. C. and Berger, J. A. (1988). Early stages of mine soil genesis in a South West Virginia spoil litho sequence. *Soil Science of American Journal*, 52: 716–723.

Robertson, M. G., Thomas, S. S. and van Zyl, D. (1980). *Uraniuil Tailings Impoundiment- Site Selection*. Symposiun on Uranium Mill Tailings Management, Geotechnical Engineering Program. Civil Engineering Department, Colorado State University, Fort Collins, Colorado, November 24–25.

Robinson, P., Hector, A., Luis, J., Benavides, D. and Hancock, D. (1979). *Uranium Mining and Milling: A Primer. The Workbook*. Vol. 4. Albuquerque, NM: Southwest Research and Information Center, pp. 6–7.

Robson, T. M., Klem, K., Urban, O., & Jansen, M. A. (2015). Re-interpreting plant morphological responses to UV-B radiation. *Plant, cell & environment*, 38(5), 856–866.

Rodrigues, B. F. (1997). Survey for potential tree species for revegetation of iron ore mine wastelands of Goa. *Indian Forester*, 123.

Rodriguez-Freire, L., DeVore, C. L., El Hayek, E., Berti, D., Ali, A. M. S., Pacheco, J. S. L. and Cerrato, J. M. (2021). Emerging investigator series: Entrapment of uranium—phosphorus nanocrystals inside root cells of Tamarix plants from a mine waste site. *Environmental Science: Processes and Impacts*, 23(1): 73–85.

Romney, E. M., Wallace, A., Gilbert, R. O., Bamberg, S. A., Childress, J. D., Kinnear, J. E. and Ackerman, T. L. (1974). Some ecological attributes and plutonium contents of perennial vegetation in area 13. In Dunaway, P. P. and White, M. G. (eds) *The Dynamics of Plutonium in Desert Environments* (USAEC Rep. NVO-142). Aberdeen Proving Ground, MD: US Army Environmental Command, pp. 91–106.

Royal Society, USA. (2000). *The Health Hazard of Depleted Uranium*. Royal Society, USA Munitions Part II, The Royal Society London, pp. 16–26.

Rufykiri, Q., Decl, E. S., Wang, L. and Thiry, Y. (2001). *Arbuscular International Ecological Engineering Society*. Switzerland, Symposium No. 64, Paper No. 2192.

Rumble, M. A. and Bjugstad, A. J. (1986). Uranium and radium concentrations in plant growing on uranium mill tailings in South Dakota. *Reclamation and Revegetation Research*, 4: 271–277. www.fs.fed.us/rm/sd/uranium_radium.Soil Technology.

Ryan, P. (1987). Rum jungle mines rehabilitation—Northern Territory. *Journal of Soil Conservation New South Wales*, 43: 18–27. Proceedings of Conference "reclamation, A Global Perspective", Calgary, Alberta, August 27–31, 1989.

Salbu, B. and Oughton, D. H. (1995). Processes affecting the uptake of radioactive species into the environment. *Radiation Protection Dosimetry*, 62: 1–4 (Abstract).

Sarangi, A. K. (2003). Grade control in Jaduguda uranium mine, Jharkhand: Transitions of the mining. *Geological & Metallurgical Institute of India (MGMI)*, 99(1–2): 2002–2003.

Sarangi, A. K., Nair, R. N., Gurunadha Rao, V. V. S., Mohanty, A. K., Ramesh, G., Dhakate, B., Tripathy, Surinaidu, L., Swaroop, Thama Rao, G., Mahesh, J., Ramesh, G., Kusmakiranmayi, K. and Prakash, B. A. (2009). *Assessment of Impact Of Uranium Tailings Ponds Around Jaduguda, Jharkhand, India*. Joint International Convention of 8th IAHS Scientific Assembly and 37th IAH Congress Water: A Vital Resource Under Stress—How Science Can Help, Hyderabad, India, September 6–12.

Sarjan, S., Jha, V. N., Sethy, N. K., Rout, S., Ravi, P. M., Jha, S. K., Tripathi, R. M. and Kulkarni, M. S. (eds). (2018). *Physico Chemical Characterization of Mill Tailings and Speciation Studies of Uranium a Jaduguda, India. India: Indian Association for Radiation Protection*. Proceedings of the Thirty-Third IARP International Conference on Developments Towards Improvement of Radiological Surveillance at Nuclear Facilities and Environment: Book of Abstracts.

Sasmaz, M., Obek, E. and Sasmaz, A. (2016). Bioaccumulation of uranium and thorium by Lemna minor and Lemnagibba in Pb-Zn-Ag tailing water. *Bulletin of Environmental Contamination and Toxicology*, 97(6): 832–837.

Sastry, C. S. T. and Kavathekar, Y. Y. (1990). *Plants for Reclamation of Wastelands*. New Delhi: Publications and Information Directorate, pp. 317–318.

Schafer, W. M., Nielsen, G. A. and Nettletan, W. (1980). Mine soil genesis and morphology in a spoil chrono- sequence in montana. *Soil Science Society of American Journal*, 44(4): 802–807.

Schlesiger, W. H. (1985). Changes in soil carbon storage and associated properties with disturbance and recovery. In Reichle, D. E. (ed) *The Global Carbon Cycle*. New York: Springer-Verlag.

Schnoor, J. L., Licht, L. A., McCutcheon, S. C., Wolfe, N. L. and Carriera, L. H. (1995). Phytoremediation of organic and nutrient contaminants. *Environmental Science and Technology*, 29: 318A–323A.

Secor, J. B., Shamash, S., Smeal, D. and Gennaro, A. L. (1983). Soil characteristics of two desert plant community types that occur in the Los Medanos area of Southeastern New Mexico. *Soil Science*, 136(3): 133–144.

Sellers, G., Hutchings, T. R. and Moffat, A. J. (2004). Remediated materials—their potential use in urban greening. *Seesoil*, 16: 47–59.

Sheppard, M. I. and Sheppard, S. C. (1985). The plant concentration ratio concept as applied to natural uranium. *Health Physics*, 48: 494–500.

Sheppard, S. C. and Evenden, W. G. (1983). The assumption of linearity in soil and plant concentration ratio for uranium: An experimental evaluation. *Journal of Environmental Radioactivity*, 7: 221–247.

Sheppard, S. C. and Evenden, W. G. (1988). Critical compilation and review of plant/soil concentration ratios for uranium, thorium and lead. *Journal of Environmental Radioactivity*, 8(3): 255–285.

Sheppard, S. C., Evenden, W. G. and Anderson, A. J. (1992). Multiple assays of uranium toxicity in soil. *Environmental Toxicology and Water Quality*, 7(3): 275–294.

Shuaibu, U. O. A. and Nasiru, A.-S. (2011). Phytoremediation of trace metals in Shadawanka stream of Bauchi metropolis, Nigeria. *Universal Journal of Environmental Research and Technology*, 1(2): 176–181.

Singh, A. (2005a). Biomass production in response to NPK fertilization in a tropical herbaceous legume Stylosanthes Hamata sown for revegetation of coal mine spoil. *Indian Forester*, 131(1): 124–126.

Singh, A. (2005b). Influence of variation in site fertility on nitrogen and phosphorus resorption efficiency in young native Tropical woody species planted on mine spoil. *Indian Forester*, 131.

Singh, B. and Raman, S. S. (1979). Physico-chemical properties of soils under Pinus petula in Eastern Himalayas. *Indian Forester*, 105(6): 482–490.

Singh, J. S. and Gupta, S. R. (1983). Plant decomposition and soil respiration in terrestrial ecosystems. *The Botanical Review*, 43(4): 449–528.

Singh, J. S., Singh, K. P. and Jha, A. K. (1995). *Integrated Ecological Studies on Revegetation on Mine Spoil. Concepts and Research Highlights*. An interim report of a S and T project sponsored by Ministry of coal, Govt. of India through CMPDT, Ranchi, Department of Botany, Banaras Hindu University, Varanasi, p. 67.

Singh, K. L., Sudhakar, G., Swaminathan, S. K. and Rao, C. M. (2015). Identification of elite native plants species for phytoaccumulation and remediation of major contaminants in uranium tailing ponds and its affected area. *Environment, Development and Sustainability*, 17(1): 57–81.

Singh, L. and Soni, P. (2009). *Species Selection for Revegetation and Consolidation of Uranium Tailings at Jaduguda*. Written by Administrator Sunday.

Singh, L. and Soni, P. (2010a). Binding capacity and root penetration of seven species selected for revegetation of uranium tailings at Jaduguda in Jharkhand, India. *Current Science*, 99(4): 507–513.

Singh, L. and Soni, P. (2010b). Concentration of radionuclides in uranium tailings and its uptake by plants at Jaduguda, Jharkhand, India. *Current Science*, 98(1): 37–49.

Singh, L., Soni, P. and Babu, T. M. (2021). *Ecological Amendment of Uranium Mine Tailings Using Native Plant Species*. In *Phytorestoration of Abandoned Mining and Oil Drilling Sites* (pp. 249–261): First Edition. Publisher: Elsevier

Singh, R. P. and Gupta, M. K. (1983). Root studies and soil binding capacity of some important grasses of temperate region (Kyaright, HP.). *Indian Journal of Soil Conservation*, 11(1): 60–63.

Singh, S., Jha, V. N., Sethy, N. K., Rout, S., Ravi, P. M., Jha, S. K., Tripathi, R. M. and Kulkarni, M. S. (eds). (2018). *Physico Chemical Characterization of Mill Tailings and Speciation Studies of Uranium a Jaduguda, India. India: Indian Association for Radiation Protection*. Proceedings of the Thirty-Third IARP International Conference on Developments Towards Improvement of Radiological Surveillance at Nuclear Facilities and Environment: Book of Abstracts.

Skeffington, R. A. and Bradshaw, A. D. (1981). Nitrogen accumulation in kaolin mining wastes in Cornwall. IV. Sward quality and the development of a nitrogen cycle. *Plant and Soil*, 62: 439–451.

Skwarzek, B. (2000). A comparison of 210- Polonium accumulation in molluscus from the southern Baltic, the coast of Spitsbergen and Sasek Wielki Lake in Poland. *Journal of Environmental Radioactivity*, 49: 201–208.

Skwarzek, B. and Falkowski, L. (1988). Accumulation of Polonium-210, in Baltic invertebrates. *Journal of Environmental Radioactivity*, 8: 99–109.

Smith, S. E. and Read, D. J. (2008). Mineral nutrition, toxic element accumulation and water relations of arbuscular mycorrhizal plants. *Mycorrhizal Symbiosis*, 3: 145–148.

Smodiš, B., Černe, M., Jaćimović, R. and Benedik, L. (2015). Transfer of uranium and radium to Chinese cabbage from soil containing elevated levels of natural radionuclides. *Journal of Radioanalytical and Nuclear Chemistry*, 306: 685–694.

Soil Survey Staff. (1951). *Soil Survey Manual, Hand Book No. 18*. Washington, DC: USDA.

Soni, P. (1995). Biorestoration of surface mined lands. *Van Vigyan*, 33: 40–52.

Soni, P. and Singh, L. (2008). *Landscape Restoration in Mined Areas -an Ecological Approach*. Homi Bhabha Centenary, DAE-BRNS National Symposium on "Landscaping for Sustainable Environment", BARC, Mumbai, IL-11, pp. 78–92, November 20–21.

Soni, P. and Singh, L. (2011). Ecotechnological approach for consolidation of uranium tailings. *Journal of Environmental Engineering and Science*, 53(3): 355–364.

Soni, P., Vasistha, H. B. and Kumar, O. (1988). Environmental conservation in mined ecosystems. In Rajwar, G. S. (ed) *Researches in Ecology, Environment Pollution*. Vol. 6. New Delhi: Today & Tomorrow's Printer & Publishers, pp. 283–297.

Soni, P., Vasistha, H. B. and Kumar, O. (1989). Ecological approach towards reclaiming mined ecosystems. *Indian Forester*, 115(12): 875–883.

Soni, P., Vasistha, H. B. and Kumar, O. (1992). *Surface Mined Lands: Problems and Prospectus*. Report on Eco-Restoration of Rock Phosphate Mined Area. ICFRE Publication-18, p. 30.

Soni, P., Vasistha, H. B. and Kumar, O. (1994). Use of non-conventional MFP species in mined land restoration. *Journal of Non Timber Forest Products*, 1(3–4): 222–225.

Šourková M., Frouz J., Fettweis U., Bens O., Hüttl R., and Šantrůčková H. 2005. Soil development and properties of microbial biomass succession in reclaimed post mining sites near Sokolov (Czech Republic) and near Cottbus (Germany). Geoderma, 129: 73–80.

Sreedhar, Y., Nagaraju, A., Thejaswi, A. and Bhargava, Y. (2017). Reclamation of abandoned mine land: A case study from Tummalapalle uranium mining area, Cuddapah District, Andhra Pradesh, South India. *Fresenius Environmental Bulletin*, 26: 1373–1382.

Srivastava, R. R., Pathak, P. and Perween, M. (2020). Environmental and health impact due to uranium mining. In Gupta, D. and Walther, C. (eds) *Uranium in Plants and the Environment: Radionuclides and Heavy Metals in the Environment*. Cham: Springer. https://doi.org/10.1007/978-3-030-14961-1_3.

Stanturf, J. A., Callaham, M. A., Jr. and Madsen, P. (2021). Soils are fundamental to landscape restoration. In John A. Stanturf, Mac Callaham (eds.), *Soils and Landscape Restoration*. Academic Press, pp. 1–37.

Stevenson, F. J. (1965). 'Origin and Distribution of Nitrogen in the Soil', in W. V. Bartholemew and F. E. Clark, (eds.), *Soil Nitrogen, American Society of Agronomy*, Madison, WI.

Sun, D., Yang, H., Guan, D., Yang, M., Wu, J., Yuan, F., Jin, C., Wang, A. and Zhang, Y. (2018). The effects of land use change on soil infiltration capacity in China: A meta-analysis. *Science of the Total Environment*, 626: 1394–1401.

Suresh, B. and Ravishankar, G. A. (2004). Phytoremediation-A novel and promising approach for environmental clean up. *Critical Reviews in Biotechnology*, 24(2): 1–28.

Svoboda, J., & Taylor, H. W. (1979). Persistence of Cesium-137 in arctic lichens, dry as integrifolia, and lake sediments. *Arctic and Alpine Research*, 11(1): 95–108.

Tabatabai, M. A. (1985). Effect of acid rain on soils. *CRC Crit Rev Environ Control*, 15: 65–110.

Tanpibal, V. and Sahunalu, P. (1989). Characteristics and management of tin mine tailings in Thailand. *Soil Technology*, 2(10): 17–26.

Thakur, C. (1957). Grassland in soil conservation. *Journal of Soil and Water Conservation*, 6(1): 43–47.

Thomas, P. A. and Gates, E. (1999). Radionuclides in the Lichen-Caribou-human food chain near uranium mining operations in Northern Saskatchewan. *Canada Environmental Health Perspectives*, 107(7): 527–537.

Tiedemann, A. R.and Klemmedson, J. O. (1973). Nutrient availability in desert grassland soils under mesquite (Prosopis juliflora) trees and adjacent open areas. *Soil Science Society of America Journal*, 37(1): 107–111.

Tordoff, G. M., Baker, A. J. M. and Willis, A. J. (2000). Current approaches to the revegetation and reclamation of metalliferous mine wastes. *Chemosphere.*

Tracy, B. L., Prantl, F. A. and Quinin, J. M. (1983). Transfer of ^{226}Ra, ^{210}Po and uranium from soil to garden produce: Assessment of risk. *Health Physics*, 44(5): 469–477.

Troeh, F. R. and Thompson, L. M. (1993). *Soils and Soil Fertility.* 5th ed. Oxford: Oxford University Press, pp. 37–87, 462.

Tsivoglue, E. C. (1964). *Environmental Monitoring in the Vicinity of Uranium Mill.* Proceedings of IAEA Symposium on Radiological Health and Safety in Mining and Milling of Nuclear Materials 1963. Vol. 2, Vienna, pp. 231–245.

Tsivoglue, E. C. and O'Connel, R. L. (1963). *Waste Guide for the Uranium Milling Industry.* Cincinnati, OH: U.S. Public Health Services, R.A. Traft Sanitary Engineering Center.

UMTRA. (2013). *Uranium Mill Tailings Radiation Control Act Sites Fact Sheet.* Albuquerque, NM: Office of Legacy Management Project Office, DOE (US Department of Energy).

UMTRCA. (1978). *Uranium Mill Tailings Radiation Control Act of 1978 Enacted by the 95th United States CongressR.* http://pubs.er.usgs.gov/publication/70178268.

UMTRCA. (2017). *UMTRCA Disposal Cell Site Selection Process Doc. No. Sl5077.* US Department of Energy. https://www.energy.gov › prod › files › 2017/01

United States Department of Energy. (1988). *Uranium Mill Tailings Remedial Action Project (UMTRAP).* Albuquerque, New Mexico, USA. pp 106–164.

UNSCEAR (United Nations Scientific Committee on the Effects of Atomic Radiation). (2000). *Sources and Effects of Ionizing Radiation.* Report to General Assembly with Scientific Annexes. UNITED NATIONS, New York.

UNSCEAR (United Nations Scientific Committee on the Effects of Atomic Radiation). (2010). *Sources and Effects of Ionizing Radiation: Exposures of the Public and Workers from Various Sources of Radiation.* New York: United Nation.

USGS. (2016). *Living Organisms Take Up Uranium from Their Environment by David Phillip.* Office of Communications and Publishing Reston, United States.

USNRC. (2020). *Protecting People and the Environment Backgrounder on Uranium Mill Tailings.* USNRC Report. U.S. Nuclear Regulatory Commission Washington, DC 20555-0001.

Varela, C., Vazquez, C., Sangregorio, G. V. M., Leiros, C. M. and Sotres, G. F. (1993). Chemical and physical properties of open cast lignite mine soils. *Soil Science*, 156(3): 193–204.

Vasistha, H. B. (1992). *Growth Behaviour of Some Colonizing Plant Species of Rock Phosphate Mine Spoils Areas of Doon Valley* (Ph.D. thesis submitted to H.N.B. Garhwal University Srinagar, Garhwal).

Vasistha, H. B. and Negi, M. (2005). Plant species colonization on abandoned limestone mined areas in Western Himalaya. *Bulletin of the National Institute of Ecology*, 16: 51–58.

Vasistha, H. B., Soni, P. and Kumar, O. (1996). *Buddleja asiatica*- a native species for restoration of degraded mined ecosystem. *Indian Journal, Soil Cons*, 24(2): 121–124.

Venkatraman, S. (1981). *Studies on Species Variation of Manganese in Uranium Processing and Natural Environment* (MSc thesis, University of Bombay), pp. 10–11.

Venu Babu, P. (2007). *Study of Vetiver for Uptake of Uranium from Tailings.* Unpublished Personal Communication.

Vera, F., Tome, M. P., Blanco, R. and Lozano, J. C. (2003). Soil-to-plant transfer factors for natural radionuclides and stable elements in a Mediterranean area. *Journal of Environmental Radioactivity*, 65: 161–175.

Vermeulen, R., Silverman, D. T., Garshick, E., Vlaanderen, J., Portengen, L. and Steenland, K. (2014). Exposure- response estimates for diesel engine exhaust and lung cancer mortality based on data from three occupational cohorts. *Environmental Health Perspectives*, 122(2): 172–177.

Vidali, M. (2001). Bioremediation: An overview. *Pure and Applied Chemistry*, 73(7): 1163–1172.

Waggitt, P. W. (1994). *A Review of Worldwide Practices for Disposal of Uranium Mill Tailings.* Australian Government Publ. Service.

Wali, M. (1987). The structure dynamics and rehabilitation of drastically disturbed ecosystems. In Khoshoo, T. N. (ed) *Perspectives in Environmental Management.* New Delhi: Oxford Publications, pp. 163–183.

Wang, W. Q., Brackhage, C., Bäuker, E. and Dudel, E. G. (2015). Rhizofiltration of U by plant root surfaces in a tailing wetland. In *Uranium-Past and Future Challenges.* Editor: Broder Merkel, Alireza Arab: Springer, pp. 497–508.

Wardle, D. A., Yeates, G. W., Williamson, W. M., Bonner, K. I. and Barker, G. M. (2004). Linking aboveground and belowground communities: The indirect influence of aphid species identity and diversity on a three trophic level soil food web. *Oikos*, 107(2): 283–294.

Watanabe, M. E. (1997). Phytoremediation of the brink of commercialization. *Environmental Science and Technology*, 31: 182–186.

Webb, J. H. and Voorhees, L. D. (1984). Revegetation of uranium mill tailings sites. *Nuclear Safety*, 25: 668–675.

White, M. G. and Dunaway, P. B. (1975). *The Radioecology of Plutonium and Other Transuranics in Desert Environments.* USERDA Rep. NVO-153. US Energy Research and Development Administration. Las Vegas, NV: Nevada Operations Office.

WHO. (1993). *World Development Report Investing in Health.* Vol. 1. World Health Organization. Washington, D.C.

WHO. (1996). *Guidelines for Drinking Water Quality.* 2nd ed. Vol. 2. Geneva: Health Criteria and Other Supporting Formation, WHO. Washington, D.C.

WHO. (2009). *WHO Handbook on Indoor Radon: A Public Health Perspective.* World Health Organization. Washington, D.C.

Wieder, R. K., Carrel, J. E., Rapp, J. K. and Kucera, C. L. (1983). Decomposition of fescae (Festuca elator var. arundinaceae) and Cellulose litter on surface mines and tallgrass prairies in central Missouri, U.S.A. *Journal of Applied Ecology*, 20: 303–321.

Wikipedia. (2012). *The Creative Commons Attribution-Share Alike License.*

Wikipedia. (2022). *Peak Uranium From Wikipedia, the Free Encyclopedia Creative Commons Attribution-Share Alike License 3.0.*

Wikipedia. (2022). Uranium. In *Wikipedia.* https://en.wikipedia.org/wiki/Uranium.

Williams, D. J. and Currey, N. A. (2002). Engineering closure of an open pit gold operation in a semi-arid climate. *International Journal of Surface Mining, Reclamation and Environment*, 16(4): 270–288.

Wilson, J. B., Wells, T. C. E., Trueman, J. C., Jones, G., Atkinson, M. D., Crawley, M. J., Dodd, M. E. and Silverttown, J. (1996). Are there assembly rules for plant species abundance an investigation in relation to soil sources and successional trends? *Journal of Ecology*, 84: 527–538.

Winde, F., Ewald, E. and Gerhard, G. (2017). Uranium contaminated drinking water linked to leukaemia—revisiting a case study from South Africa taking alternative exposure pathways into account. *Science of the Total Environment*, 574: 400–421.

Woodmansee, R. G., Reeder, J. D. and Berg, W. A. (1980). Nitrogen in drastically disturbed lands. In Youngberg, C. T. (ed) *Forest Soils and Land Uses*. Fort Collins: Colorado State University, pp. 376–392.

World Nuclear Association. (2008). *Supply of Uranium*. http://www.world-nuclear.org/info/inf75.htm

World Nuclear Association. (2021). *Nuclear's Contribution to Achieving the UN Sustainable Development Goals Registered in England and Wales*. Company Number 01215741.

Xiaodong, C. P., MacNaughton, J., Cedeno, L. and Joseph, G. A. (2017). Radon- induced lung cancer deaths may be overestimated due to failure to account for confounding by exposure to diesel engine exhaust in BEIR VI miner studies. *PLoS One*. 12(9): e0184298. https://doi.org/10.1371/journal.pone.0184298

Yan, D., Ding, K., He, Y., Fan, L., Zuo, R. and Jiang, X. (2018). Uptake of uranium by plants growing around an old uranium mill tailing in northeast China. *Fresenius Environmental Bulletin*, 27: 3102.

Yan, X. (2016). Uptake of radionuclide thorium by twelve native plants grown in uranium mill tailings soils from South part of China. *Nuclear Engineering and Design*, 304: 80–83.

Yang, B., Zhou, Q. and Zhang, J. (2021). Evaluation of the natural radioactivity in food and soil around uranium mining region. *Journal of Radioanalytical and Nuclear Chemistry*, 329: 127–133. https://doi.org/10.1007/s10967-021-07794-w.

Yang, X., Feng, Y., He, Z. and Stoffella, P. J. (2005). Molecular mechanisms of heavy metal hyperaccumulation and phytoremediation. *Journal of Trace Elements in Medicine and Biology*, 18: 339–353.

Yang, X. E., Long, X. X., Ni, W. Z. and Fu, C. X. (2002). S*edum alfredii* H: A new Zn hyper accumulating plant fist found in China. *Chinese Science Bulletin*, 47(19): 1634–1637.

Yuarsah, I., Handayani, E. P., Rakhmiati and Yatmin. (2017). Restoration of soil physical and chemical properties of abandoned Tin- Mining in Bangka Belitung Islands. *Journal of Tropical Soils*, 22(1): 21–28. http://journal.unila.ac.id/index.php/tropicalsoil. DOI: 10.5400/jts.2017.22.1.21.

Zhuang, P., Yang, Q. W., Wang, H. B. and Shu, W. (2007). Phytoextraction of heavy metals by eight plant species in the field. *Water, Air, & Soil Pollution*, Springer Nature, 184: 235–242. https://doi.org/10.1007/s11270-007-9412-2.

Zinke, P. J., Strangenberger, A. G., Post, W. M., Ennuel, W. R. and Olson, J. S. (1984). *Worldwide Organic Soil Carbon and Nitrogen Data* (Pub. No. 2212 (DE 84012041) (ORNL/TM-8857)). Oak Ridge, TN: Environmental Sciences Division, National Laboratory.

Index

For Product Safety Concerns and Information please contact our EU
representative GPSR@taylorandfrancis.com
Taylor & Francis Verlag GmbH, Kaufingerstraße 24, 80331 München, Germany